Ocean Recovery

OCEAN RECOVERY

A Sustainable Future for Global Fisheries?

RAY HILBORN AND
ULRIKE HILBORN

OXFORD
UNIVERSITY PRESS

OXFORD

UNIVERSITY PRESS

Great Clarendon Street, Oxford, OX2 6DP,
United Kingdom

Oxford University Press is a department of the University of Oxford.
It furthers the University's objective of excellence in research, scholarship,
and education by publishing worldwide. Oxford is a registered trade mark of
Oxford University Press in the UK and in certain other countries

© Ray Hilborn & Ulrike Hilborn 2019

The moral rights of the authors have been asserted

First Edition published in 2019

Impression: 2

Published in the United States of America by Oxford University Press
198 Madison Avenue, New York, NY 10016, United States of America

British Library Cataloguing in Publication Data

Data available

Library of Congress Control Number: 2018967565

ISBN 978-0-19-883976-7

DOI: 10.1093/oso/9780198839767.001.0001

Printed and bound by
CPI Group (UK) Ltd, Croydon, CR0 4YY

PREFACE

Why write a book about world fisheries? Because fisheries are vitally important to global food security.

But, by their very nature—fishing boats are far out at sea—their importance is mostly unnoticed or worse, their very existence is being challenged and portrayed as evil. For several decades now, the public narrative, particularly about marine fisheries, has been to vilify ocean fishing with half-truths, plain-old lies, and more than a bucketful of misguided emotion. In light of the ever-increasing population of our world, it seems downright reckless to use political pressure via public misinformation to potentially reduce everyone's food supply.

More than 40 million people are professionally catching and processing fish, and more than 1 billion of the poorest people on earth depend on fish protein for a substantial part of their diet. Global capture fisheries produce more food than global beef production does. When we add the badly documented freshwater fisheries, capture-fish production may well exceed global pork and poultry. The world is eating a lot of fish.

And yet. Fisheries collapsing and species going extinct right and left have been the stories the media and much of the scientific literature, even a stack of books, have frightened us with headlines like "all fish stocks will be collapsed by 2048," or "all the large fish of the ocean are gone," and long articles in *Time* and *National Geographic* convinced us of the demise of world fisheries. Sustained non-governmental organization (NGO) campaigns lament the disappearing fish, ask for donations, and advise us on what and what not to eat to save the oceans.

Unnoticed in all this frenzy of doom, a stubbornly persistent cornucopia of offerings at the fish counters of the Western world has remained the same, has possibly increased even. There has been no falling off in bounty.

It should come as no surprise that the real situation is complex: there are failures and successes. Even though there have been no fish extinctions, there have been major fish stock collapses, but there have also been recoveries. The ocean, far from being drained of them, still contains lots of fish, perhaps as many as there were before industrial fishing began. It is almost perverse that it is precisely in most of the doom-laden developed countries that fish stocks are not even the tiniest bit declining, but increasing. The disconnect between the well-stocked seafood counters and the common narrative of gloom and doom is truly perplexing.

I recognize that most people find themselves outside the world of fisheries management and are not familiar with the great complexities of past and present developments. I also recognize that the ever-growing pressure from advocacy groups can easily take us in the wrong direction, away from solutions that have been shown to work. Hence, this book.

My aim is to provide interested readers outside of fisheries science with an overview of the major issues associated with managing global fisheries, but also with enough detail that anyone can dig into specific topics. My hope is to correct many of the most common misconceptions about how fisheries are managed, the status and sustainability of fish stocks, and the relative environmental cost of catching fish in the ocean compared with producing food from the land.

Some teasers that may surprise you and encourage you to read more.

In most of the developed world, the abundance of fish is increasing not decreasing. Overall, global fish abundance is probably stable.

The number of fish in the ocean now is likely higher than it was before industrial fishing began, and their total weight may be close to what it was before large-scale fishing. (Hint: Many fisheries target predators; therefore, more prey survive. Smaller fish in greater numbers equals more total weight.)

Overfishing can be sustainable.

There is no widespread extinction of marine fish. Only one marine fish has been documented to go extinct, and it was not due to fishing.

Marine Protected Areas (MPAs) will only increase the abundance of fish where fish stocks are seriously overfished.

Commercial fishermen have more of a financial interest in sustainability than NGOs or academics like me. A collapse puts fishermen out of business, whereas NGOs and academics see their funding increase.

Capture fisheries are among the lowest environmental impact forms of food production. A thoughtful piscivore could have a lower environmental impact than that of a vegan.

This book is the result of more than 40 years working on fisheries around the world and in the last 10 years, working with a number of international teams on some of the most contentious topics in fisheries. In particular, my work on the global status of fisheries began with a study group in 2008–2009, jointly led with Prof. Boris Worm of Dalhousie University. Many of the same group met again during the last few years to update the data and understanding of the status of fisheries. Other study groups have been looking at the global impact of bottom trawling, fishing for forage fish, and marine protected areas. I have drawn heavily on their results.

The book is written in the first person, although it is coauthored. The first draft of each chapter's basic content was written by Ray Hilborn. Ulrike Hilborn then rewrote each chapter to make it more accessible, eliminate the jargon Ray cannot seem to avoid, and make the intent clearer. In many cases, Ulrike drew on her own experience, including anecdotes and examples.

ACKNOWLEDGMENTS

This work is the product of 40 years of collaboration with a wide range of people. The most influential have been Ricky Amoroso, Trevor Branch, Chris Costello, Marc Mangel, Mike Melnychuk, Ana Parma, André Punt, Tom Quinn, Daniel Schindler, Paul Starr, Kevin Stokes, and Carl Walters.

The Food and Agriculture Organization of the United Nations is the global resource on fisheries and its staff members are among the world's most knowledgeable people about fisheries. I especially thank current and former staff including Arni Mathiesen, Manuel Barange, Yimin Ye, Petri Suuronen, Alejandro Anganuzzi, Jessica Sanders, and Nicolas Gutierrez for inspiration, information, and guidance.

One of the best parts of my job is the interaction with students and postdoctoral fellows, and I have been fortunate in having had a wonderful and creative group to work with.

Any discussion of fisheries depends entirely on the thousands who collect the data, attend meetings, and perform the analyses that I have used.

Many people have provided data, examples, and fact checking of this book. I thank Dean Adams, Milo Adkison, Robert Arlinghaus, Manuel Barange, Trevor Branch, Joe Bundrant, Doug Butterworth, Al Chafee, Kevern Cochrane, Karl DeLong, Miriam Fernandez, Frances Gulland, Steve Hall, Anne Hilborn, Gordon Holtgrieve, Bill Karp, Mike Kaiser, Michael Leonard, Kai Lorenzen, John Lowrance, Mike Melnychuk, Pamela Mace, David Middleton, Dave Moore, Mike Nussman, Mike Orbach, Ana Parma, Richard Parrish, Ernesto Peñas, Joe Plesha, Jake Rice, Brian Rothschild, Lisa Seeb, Ian Sherman, Paul Starr, Daryl Sykes, and Norman Van Vactor.

Information and drawings for figures were supplied by Ricky Amoroso, Nicole Baker, Tyler Dann, Roland Pitcher, John Olson, and Cathy Schwartz.

CONTENTS

CHAPTER 1

The Bristol Bay Salmon Fishery

July 3, 2017, was a very good day for Nick Lee and his crew in the *Anasazi*, a 32-foot gillnet boat fishing in the Nushagak District of Bristol Bay Alaska. They caught 9,000 sockeye salmon, worth roughly $60,000. In fact, July 3 was a very good day for almost everyone fishing in the Nushagak—a record 1.3 million fish were caught by the 545 boats fishing there. For some, the fishing was too good—four boats sank, overloaded with their bounty. Fortunately, no lives were lost, but those four boats missed out on an exciting season.

2017 was the third year in a row when 50 million sockeye salmon returned to Bristol Bay in a very short summer season. Each year, the catch exceeded 37 million fish. 18.7 million fish evaded the fishing nets and swam upriver to spawn, well in excess of the number estimated to produce the maximum long-term harvest. Despite the record catches, there were still plenty of fish on the spawning grounds to assure the future of the run. Bears, eagles, ravens, gulls, terns, rainbow trout, char, and the tiny shrews that depend on the salmon run also had a very good year. Surprisingly, despite the high volume of fish, the price did not fall as it usually does in superabundant years, creating that unusual circumstance for fishermen and processors where lots of fish fetched a good price. The total value of Bristol Bay sockeye landed in 2017 was $209 million. Happiness all around.

At the end of the season, flying back to Seattle from Bristol Bay, my wife sat next to a man with a badly bandaged finger. Asked if he had hurt himself fishing, he said no, but in a rush to put his gillnet boat and gear away he had smashed his finger and, pressed to catch his flight and for want of anything better, had put duct tape over the wound. "How was the season?" she asked, and he leaned back and sighed, "Epic." Later that year I phoned Nick Lee and told him about the fisherman on the plane, asking if his season too had been epic. Nick said, "Well I caught more fish than I ever have before, and I made more money than ever before, so yes, I think it was epic."

Ocean Recovery: a sustainable future for global fisheries? Ray Hilborn and Ulrike Hilborn, Oxford University Press (2019). © Ray Hilborn and Ulrike Hilborn 2019. DOI: 10.1093/oso/9780198839767.001.0001

Between 1950 and 2015, sockeye salmon have been worth 9.7 billion dollars to US fishermen, exceed only by Maine lobster, worth 10.3 billion dollars. Most sockeye caught in the USA come from Bristol Bay, a fishery that is often described as one of the most sustainable in the world, which makes it a good place to start exploring the sustainability of world fisheries.

In the summer of 1983, Chuck Meacham, then the head of research for the Commercial Fisheries Division of the Alaska Department of Fish and Game, invited me to come and see Bristol Bay. From the air, I saw the green and blue wilderness below, the lakes, mountains, and remote rivers that have sustained the local communities for thousands of years and provided so much salmon to feed the world.

The Mighty Sockeye

Sockeye salmon are unique among the Pacific salmon because almost all depend on large lakes for their first one or two winters after they hatch from eggs. Bristol Bay has lots of the right kind of lakes, including Lake Iliamna, the largest US lake after the Great Lakes. Flying over Bristol Bay with the staff of Fish and Game, I saw the large rivers that are the highways the fish travel to get to the lakes, the lakes themselves, and the smaller streams and rivers where the sockeye salmon also spawn. There are nine large rivers in Bristol Bay that drain more than twelve large lakes (see Figure 1.1).

The glaciers of the ice ages shaped this sockeye salmon paradise. They carved the large lake basins and left behind extensive gravel beds that are up to 10 or more feet deep. Perfect for the sockeye to make their nests, called redds, and lay their eggs. Bristol Bay has everything to keep millions of sockeye happy. Gravel of just the right size to spawn, lakes for their freshwater life, and a river that takes them to and from the Bering Sea and North Pacific Ocean, where they can grow to maturity.

Fieldwork, Data, and Forecasting

Ever since 1946, the University of Washington (UW) has maintained a series of field camps and a summer research program in Bristol Bay, studying sockeye salmon, the freshwater ecosystem, and salmon management. In 1995 I was invited to visit these camps and decided to make Bristol Bay one of my major research areas. Since then I have spent much of each summer in "the Bay," and I got to know the ecosystem and the people. During the fishery in the last week of June and the first two weeks of July, the members of the UW Alaska Salmon Program and I assemble data from the commercial fishery and from a research boat that samples the fish about 200 miles west of Bristol Bay. We provide daily estimates of how many fish are coming into Bristol Bay and where they are headed to help fishermen decide where to fish and processing companies to know what to expect.

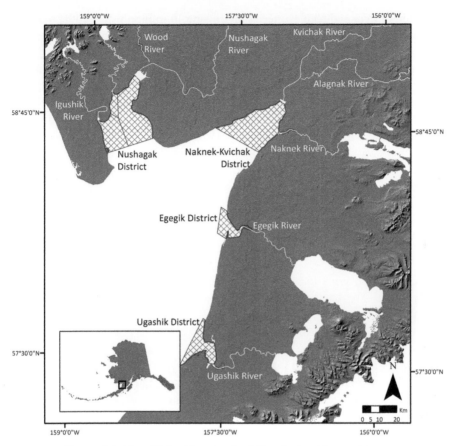

Figure 1.1 Map of Bristol Bay showing fishing districts, lakes, and rivers.

In August, when the adult fish have moved out of the lakes and into the streams to spawn, I come back to join faculty, students, and technicians in a range of studies on the fish and their impact on the ecosystem. We walk the creeks; count, tag, and measure the sockeye; take genetic samples; and see how many have been eaten by bears, eagles, ravens, or gulls.

From Canneries to High-Tech Processing: The Annual Gamble on Sockeye Returns

Salmon recolonized the watersheds of Bristol Bay after the last ice age as the glaciers retreated. Since the arrival of the first Native Americans, the salmon runs of Bristol Bay have made settlements possible, providing food all year round for people who salted and air-dried the fish. Industrialization came with the arrival of the canneries. Salmon-canning technology was developed in California after the gold rush, and once Alaska had been bought from Russia in 1867, the canneries moved north. The first cannery in Bristol Bay was established in 1883, and by

1901, there were eighteen canneries processing 10 million fish. By 1912—and for the rest of the twentieth century—an average of 20 million fish a year were canned.

But there were ups and downs. In 1919, the catch was only 7 million fish. The early 1970s brought the worst runs in history, with almost no catch allowed. Then, in the late 1970s, the runs boomed, still with ups and downs, but the average catch has been over 25 million fish since 1978 (see Figure 1.2).

A Closer Look

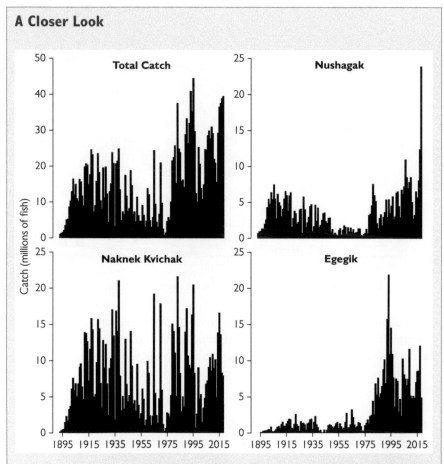

Figure 1.2 The history of catch in Bristol Bay with total catch (in millions of fish) in the upper left panel and the three other major fishing districts in the other panels. Data from the Alaska Department of Fish and Game.

Note how the Egegik district boomed after 1975, whereas Naknek Kvichak has had many ups and downs but no notable change after 1975.

Up until the early 1980s, Bristol Bay processors canned salmon—that was the business. This all changed when the USA, Canada, and Russia convinced the Japanese to stop fishing for salmon on the high seas, where they had been catching many Bristol Bay fish for the Japanese market. In return, the USA and Canada would not carry out their threat to reduce imports of Japanese cars. The Japanese, on the other hand, were not keen on canned salmon but much preferred it frozen. Eyeing such a potentially large market, companies rapidly developed freezing technology—sending fish to Japan with head and gut removed, called H&G. Prices soared, and in 1989, one sockeye salmon delivered by a fishing boat was worth over $2 per pound or $10 per fish, the same price as a barrel of oil.

But this bonanza did not last. Norway had perfected salmon farming, and as the production of farmed salmon ramped up and started to displace sockeye in markets, the price of H&G salmon declined. Worse yet, the catch from 1997–2004 averaged below 15 million fish. The price reached its nadir in 2002 at $0.45 a pound, when one sockeye bought not quite a whole big Mac, a far cry from a barrel of oil. These were terrible days in Bristol Bay. I vividly remember the 1997 season, the first poor year, with a catch of only 11.9 million fish. By the Fourth of July, it was clear that the run was a bust, and the canneries sent their summer hires home. The expected summer wages disappeared. The airport in King Salmon, where most seasonal workers fly to, was crowded with more than a thousand people trying to change their return flights, having made no money. It was equally dismal on the fishing grounds, where boats sat idle day after day, as only minimal fishing was allowed. Processing companies were bleeding red ink, and much of the fleet was in deep trouble.

Since 1975 Bristol Bay, like all Alaskan salmon fisheries, has been managed with limited entry, that is, only a certain number of permits to fish are granted. Initially these were given to individuals who could show a history of fishing in a particular area. But the permits can be sold, and by the early 1980s, they were worth $250,000. That dropped to $10,000 with the 1997–2004 decline, and anyone who had borrowed money to buy a boat or a permit in the early 1990s was likely unable to make payments on the loans. Equally hard times came to local communities that rely on the landing tax to fund their schools and public services.

However, since 2004, the runs have increased greatly, and catch has risen to an average of over 25 million fish. And so has the price. Thanks to the marketing efforts of the Alaska Seafood Marketing Institute, and especially the work of people like John Lowrance, the price went up.

Lowrance moved to Alaska in 1982 when he was 21. He worked for various fishing companies and eventually started his own, Leader Creek, in Bristol Bay, in 2000. He realized that to improve the value of Bristol Bay sockeye, he needed to produce high-quality fillets rather than H&G. Much of the fleet did not have refrigeration, but Lowrance insisted that all boats fishing for him have refrigerated seawater systems to keep the fish cold from the moment they are caught, and that the fish must be bled. He also started a profit-sharing plan for the fishermen, with forensic accountants determining each one's share.

Fillets fetch a much higher price than H&G, and for the first time, Bristol Bay sockeye were widely sold in US markets. Soon all the other processing companies followed Lowrance's lead, installing fillet lines and insisting on the immediate refrigeration of fish on board vessels. At the same time that prices rose, runs also increased, and "epic" fishing seasons happened.

A Vast Ecosystem Surrounds the Sockeye

The Bristol Bay ecosystem is far more than just sockeye salmon. There are four other species of salmon and resident fish, including rainbow trout, char, and grayling. On land, there are bears, moose, caribou, wolverines, wolves, foxes, and beavers. Many of these species depend on the annual pulse of spawning salmon that bring eggs for the resident fish and salmon bodies for the bears, eagles, ravens, and gulls. With millions of fish being taken each year, what has been the impact of industrial fishing for salmon on the ecosystem?

For the answer, we have to look at Daniel Schindler's work.

Daniel spent his childhood summers in his father's field camp in central Canada and was a champion dogsled racer. When he joined the faculty at the UW, his PhD supervisor urged me to "take Schindler to Alaska." So I did, and ever since 1999, Daniel has been spending his summers in our Bristol Bay field camp.

One of his projects has been to reconstruct the history of salmon abundance in sockeye salmon lakes all around the North Pacific for the last 500 years.[1] He takes a core of mud from the bottom of the lakes and analyzes the layers for the concentration of nitrogen isotopes that the sockeye are bringing from feeding in marine ecosystems. This allows Schindler to estimate, from the nitrogen isotopes that the decaying carcasses left behind, how many fish spawned in the lake each year (see Figure 1.3).

When industrial fishing began around 1900, the escapement, that is, the number of fish that returned to the spawning grounds, declined. However, despite the new

A Closer Look

The 300-year history of number of fish spawning in Lake Nerka (solid black line) is shown below, along with the data since 1950, when the total return to Lake Nerka can be estimated (gray line). Before 1900, the total return to the lake was roughly 1 million sockeye, but when industrial fishing began, the number reaching the lake declined, as many fish were caught and canned. When good data became available starting in 1950, the total return to Lake Nerka was about 1 million fish. This suggests that the fishery, while removing roughly half of the potential spawners, was not having a big influence on how many returned. Starting in the late 1970s, the total return began to increase considerably.

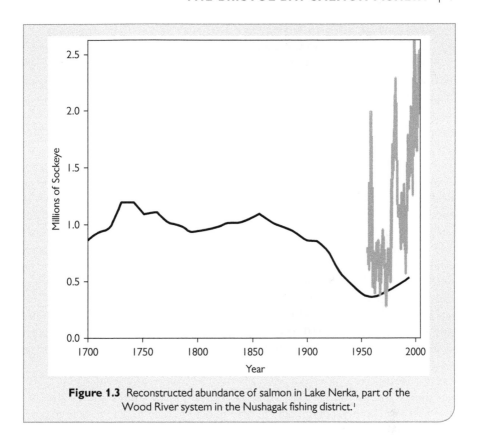

Figure 1.3 Reconstructed abundance of salmon in Lake Nerka, part of the Wood River system in the Nushagak fishing district.[1]

predator, the total number of fish returning to Bristol Bay, that is catch and escapement, kept naturally fluctuating as it had for the last several hundred years, a sign that before industrial fishing, there were definitely more than enough fish that returned to the spawning grounds. We have seen this over the years when too many sockeye escape the fishery and find to their dismay that there is no room: all of the redds are taken, and they die unspawned.

As for the sockeye's natural predators, we cannot tell if there are fewer of them now than there were before commercial fishing began. Each summer I study two small streams intensively. The field crew and I tag each fish; we record their location in the stream each day and we record their deaths: for example, were they killed by a grizzly, and if so, how much of the fish did the bear eat. Bears do not necessarily eat what they kill. Some nights they kill each fish in the stream but eat very few. Not all bears are equally skilled, and salmon have many ways of escaping the grasp of those fearsome claws. On the whole, it seems that despite the reduced escapement, there are more than enough salmon for the bears and resident fish, and the ecosystem remains largely unchanged.

How Is It that a Fishery that Is over 100 Years Old and Has Boomed for over 30 Years Has Never Collapsed?

Brief answer: management. Ever since Alaska became a state in 1959, the Alaska Department of Fish and Game has been explicitly directed by the state constitution to manage for sustainability. Management is a two-step process. The first step is an escapement goal for each river, the desired number of fish to reach their spawning grounds. The second step is the in-season daily management. The fishery in each fishing district off the mouth of each river is opened or closed daily. Counting towers on each river system allow Fish and Game staff to count the fish as they swim upstream. With a tower on both sides of the river, counting each fish is possible because the sockeye stick very close to the edges and the rivers are crystal clear. For 10 minutes of each hour, the fish passing through are counted on each side, and that number is multiplied times six. It is perhaps the most easily counted fish population of any major fishery.

The managers know how many fish have gone upstream each day, and they compare that to the target number that should have passed by the counting towers to reach the final escapement goal by the end of the season. If the count is behind the cumulative target for the day, the fishery remains closed to allow more fish to go upstream. If the count is ahead of the target, the fishery is opened. Naturally, it is a bit more complex, but that is the essence.

On the Frontline Is the Manager

The heart and soul of the management system is the management biologist who decides each day whether to open or close a fishing district. Tim Sands was the management biologist for the Nushagak District, where Nick Lee fished in 2017. Each day Tim gets the counts from the towers on the Wood River, the Nushagak River, and the Igushik River and announces on the radio where fishing will be allowed the next day. His office is in Dillingham, where fishermen can drop in and ask why he is opening or closing the fishery. In a good year like 2017, everyone is too busy to visit, but in a year of poor returns when the fleet is often tied up for days on end, there can be a long line outside his office. On the east side of the bay at the Fish and Game office in King Salmon, Travis Elison is the manager of the Naknek/Kvichak fishing district, and Paul Solomone manages the Egegik and Ugashik districts. Like Tim Sands, they make their daily decisions, and each day of the fishing season, they are available to talk to the women and men who fish in Bristol Bay.

It does not always go well. Tim Sand's predecessor once kept the fishery closed while nearly one million fish swam up the Wood River past the counting tower. That one million was almost the entire escapement needed for the year. Even though the fleet had told him that there was a great run of fish in the district, he was too concerned to make his escapement and kept the fishery closed. The fishermen in Dillingham posted a petition at the grocery store demanding his resignation.

In 2017, a similar situation threatened Tim Sands. He had exceed his escapement goal for the Nushagak River very early in the season, and fishermen thought he was being too conservative. But in the end, he kept the fishery wide open for several weeks, the fleet kept fishing; the fish kept coming; and escapement goals were met and exceeded. Everyone was happy with the season, and no demands for resignation were heard.

The direct individual responsibility of managers and personal access to them in Bristol Bay is unique to Alaska. In contrast, federal fisheries are managed by Regional Fisheries Management Councils, committees that makes decisions with diffuse individual responsibility. Again unique to Alaska is that each manager's primary job is to make the escapement goal. If it is not reached, he or she must answer to the supervisor. In other salmon fisheries outside Alaska, managers are expected to balance the conservation of the stock with providing a livelihood to the fishermen. In Bristol Bay, as in most of Alaska, conservation of the resource has the highest priority.

What Makes the Bristol Bay Salmon Fishery so Biologically Sustainable?

I believe that there are three key elements. First is the pristine habitat. This vast ecosystem has seen no significant human impact except climate change, and its lakes and streams are unaltered. And there is the *portfolio effect*[2] that encompasses the entire Bristol Bay system. Before the 1980s, the Kvichak River and Lake Iliamna dominated the fishery. The Egegik, Nushagak, and Ugashik fishing districts were insignificant, and the money was reliably in the Kvichak, where the average catch was over 7.5 million. Poor Egegik came in with fewer than 2 million fish from 1920 to 1990, and was expected to remain so, when in the early 1990s, the Egegik suddenly boomed and averaged almost 11 million fish, while Kvichak came in slightly lower. From 2000 to 2009, the Egegik again had more catch than the Kvichak did. Then, to everyone's even bigger surprise, the Nushagak surged ahead in 2006, with the highest catch in the Bay, and it repeated that feat in 2017. Salmon, it appears, are more inscrutable than we think.

How easy it would have been in the 1950s, say, to put a dam or a mine in the Egegik or Nushagak districts because the salmon and the money were elsewhere, and the potential of both river systems would have been lost forever. This is precisely what happened to many watersheds in the lower-forty-eight states.

A Closer Look

The *portfolio effect*, a term taken from the stock market, is simply a fancy version of the old adage of not putting all your eggs in one basket. A wise investor spreads money across a range of investments, so that while some go down, others may be going up, and the overall investment is much more stable than it would be if everything were in one company. The portfolio effect works best when there is a negative correlation between alternative investments—if there is a positive correlation and all possible investments go up and down together, you do not really achieve the portfolio effect.

The portfolio effect in the context of Bristol Bay is that the different river systems do not go up and down together; while the Kivchak River was doing poorly in the 1990s, Egegik was booming. There are also a range of portfolio effects within each river and the salmon themselves. Sockeye salmon return to Bristol Bay dominantly at ages 4, 5, and 6 years, so that a really successful spawning year spreads the return out over 3 different years, providing more stability than would happen if they all returned at the same age. Within each river, there is a portfolio of different habitats, typically small streams, large rivers, and lake beaches. These too seem to operate a bit out of sync, thus providing further stability.

The second salient element was the elimination of other fisheries outside the Bristol Bay management system that used to take a significant portion of the run. Closing the high seas Japanese fishery was the key step. One Alaskan fishery remains, located west of Bristol Bay, and called Area M, that captures Bristol Bay fish before they enter the fishing districts. While there are frequent battles between Bristol Bay and Area M fishermen, the Area M catch is small enough not to threaten the sustainability of the Bristol Bay fishery.

The third critical element is the culture of Alaska Department of Fish and Game and its emphasis on meeting escapement targets as the first priority. Integral to this is the science data-collection system, especially the counting towers, and a host of other systems, including sampling size and age of fish in both the catch and the escapement.

Of course, other factors contribute to sustainability, with the limited-entry system at the top of the list. There are about 1,800 permits for gill net boats in Bristol Bay, which reflects the number of boats that fished in the 1970s when the limited-entry system was established. When both runs and prices boomed in the 1980s—remember when one fish was worth one barrel of oil—without limited entry, there would probably have been 10,000 boats trying to fish Bristol Bay, a nightmare for any manager.

Clouds over Paradise

I have presented the Bristol Bay fishery as a prime example of a sustainable fishery, and without a doubt, the region is eminently productive, and since enough fish are allowed to reach the spawning grounds, it will remain so. But there are considerations other than biology. The economic sustainability is much more fragile because it depends on the cost of fishing and global prices for fish, both of which are outside the direct control of managers. A 2002 study suggests that reducing the fleet to perhaps 800 boats would provide for more economic stability, but at what cost to local participation?

The social sustainability is also in question. Originally, the majority of Bristol Bay permit holders were residents of the Bristol Bay Region. Over time, the number of residents holding a permit has gone down because some have moved to big cities such as Anchorage or have sold their permits on the open market.

A Closer Look

The Pebble Mine is a proposed development of large deposits of gold, copper, and molybdenum that lies on the border of the Nushagak and Kvichak Rivers.

It has been advertised as the second-largest deposit in the world. Under the Obama administration, the EPA was on the verge of preemptively saying that it would violate the

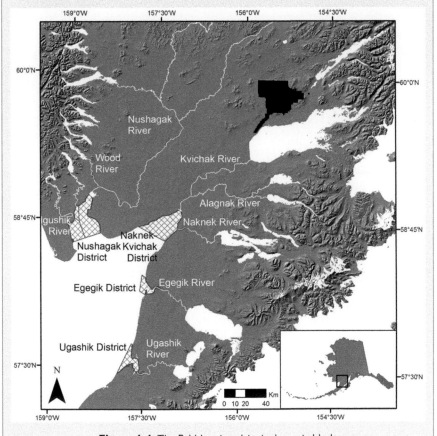

Figure 1.4 The Pebble mine claim is shown in black.

Continued

A Closer Look *Continued*

Clean Water Act, but the Trump administration reversed this, and a specific proposal is expected to be put forward for permitting. It is feared that the toxic wastewater will either leak through the same porous geology that makes the region such a great spawning area or that any dams constructed to hold the wastewater will ultimately fail.

As to the future of the fishery, there are a number of threats, foremost the proposed Pebble Mine, to be sited between the Nushagak and Kvichak watersheds (see Figure 1.4). As planned, it would be the largest open-pit copper and gold mine in the world, with its attendant highly toxic liquid waste to be stored in perpetuity behind an earthen dam higher than the Golden Gate Bridge in an earthquake zone, also prone to volcanic eruptions. Clearly, this undertaking should never have been born, yet its possibility keeps haunting Bristol Bay.

The decline in oil prices and consequent royalty revenue for the State of Alaska is a very real peril to the funding for the Fish and Game science program. The future of the counting towers and all the other data collections that are essential for the management of the fishery is highly uncertain.

Climate change is real, and Western Alaska is one of the fastest-warming places in the world. In the mid-1990s, essential field gear for our Alaska program was neoprene chest waders to endure the daily summer rain and cold. Fewer than 10 years later, we switched to lightweight Gortex, shed our many layers, and swam in the lakes! So far the sockeye too seem to like the rising temperatures. Because they are at the northern end of their range, longer and warmer summers have meant a longer growing season in the lakes. But we shall see what the future brings.

Post scriptum: All records were broken in 2018, when 62.3 million sockeye returned to Bristol Bay—the largest return in history!

FURTHER READING

A good history of Bristol Bay. Bristol Bay Economic Development Corporation. 2003. An analysis of options to restructure the Bristol Bay salmon fishery. Dillingham, Alaska. (http://www.bbsalmon.com/FinalReport.pdf).
Estimation of the history of spawning stock and the impact of the fishery for 250 years. Schindler DE, Leavitt PR, Brock CS, et al. 2005. Marine-derived nutrients, commercial fisheries, and production of salmon and lake algae in Alaska. Ecology 86: 3225–31.
A study of the loss of permits to local fishermen in Alaska. Knapp G. 2011. Local permit ownership in Alaska salmon fisheries. Mar Pol 35: 658–66.
More on the portfolio effect. Schindler DE, Hilborn R, Chasco B, et al. 2010. Population diversity and the portfolio effect in an exploited species. Nature 465: 609.

CHAPTER 2

Fisheries Sustainability

Definitions of Sustainability

In the world of fisheries, there are many stakeholders with conflicting desires. Recreational and commercial fishermen fight political battles over who gets to fish, and so do industrial and artisanal fishermen. Environmental protection groups want less fishing to protect marine ecosystems, whereas those who catch fish would like to catch more. But one thing they all agree on is they want the fishery to be sustainable. But what is *sustainability*, and what does it actually mean?

Luckily, we have a definition that is well thought out and widely accepted. It is the Brundtland Commission's report, named after Gro Harlem Brundtland, former Prime Minister of Norway. It can be found in a 1987 report called *Our Common Future*, issued by the World Commission on Energy and Development,[1] and it reads:

> Sustainable development is development that meets the needs of the present without compromising the ability of future generations to meet their own needs.

The Brundtland Commission recognized that the goal of development has to be to provide benefits to people, such as reducing poverty, improving health, and making the world a better place for them, but we must also recognize that the ability of the earth to provide the resources to produce these benefits is limited. Whatever we do now that depletes resources such that our children and children's children cannot enjoy them is not a path to a better future. The Brundtland report is a major shift away from the earlier focus on sustainability for only the natural world, to an emphasis on the connectedness of people and the natural world, recognizing that trade-offs must be made.

Major international groups, such as the International Union for the Conservation of Nature, now recognize that sustainability has three equally important elements: environmental, social, and economic sustainability. But in discussion on fisheries sustainability, the social and economic components have been hotly debated without yielding consensus.

Ocean Recovery: a sustainable future for global fisheries? Ray Hilborn and Ulrike Hilborn, Oxford University Press (2019). © Ray Hilborn and Ulrike Hilborn 2019. DOI: 10.1093/oso/9780198839767.001.0001

Sustainability has become so respectable in the science world that major journals now have specific sections and editors dedicated to the topic. One of these editors for the venerable Proceedings of the National Academy of Sciences of the United States (PNAS) is William C. Clark of Harvard University. He and I were PhD students at the University of British Columbia and colleagues at a think-tank in Austria in the 1970s, and he remains my go-to scholar on sustainability. He and two colleagues recently published *Pursuing Sustainability: A Guide to the Science and Practice*,[2] where they identify the objective of sustainability as "inclusive measures of well-being":

> Our approach is grounded in a definition of sustainable development that focuses on the well-being of people over the long run. In particular, it argues that the development of a social-environmental system is sustained over a period of time if over that period appropriately inclusive measures of well-being do not decline.

Both Brundtland and Clark see attaining sustainability in terms of managing a social-environmental system with the objective of well-being for people. In this endeavor, we are, and will always be, constrained by the limits of the natural world. If we degrade it, our well-being will diminish.

But how do we judge the sustainability of a particular activity such as a fishery? Some think it is a condition, i.e., state based, and others, that it is a process.

Those who think of sustainability as a condition will argue that if a fish stock is at high abundance, it is sustainable, and if at low abundance, it is not sustainable. Sustainable seafood guides are very much state based. Fish that are relatively abundant are okay to eat, but fish that are overfished must be avoided.

For those who think of it as a process, a fishery that is managed to provide long-term benefits is sustainable. Equally, a fish stock at low abundance, but managed to allow rebuilding it, is sustainable, but a fish stock that might be relatively abundant but is not managed and likely headed toward depletion is not sustainable.

As we will see in later chapters, both views are alive and well in the world of seafood, but I strongly support that sustainability is achieved through a process, and that state-based measures are often flawed.

Sustainable Exploitation of Fish Stocks

Fish stocks and marine ecosystems are a renewable resource: we can take fish from a population forever so long as we do not take too many and the underlying ecosystem is maintained. Since natural systems fluctuate in their productivity, we clearly must take less during the less productive times than when the system is productive.

In Chapter 1, I said that the Bristol Bay fishery is sustainable because the habitat is intact and the catch is regulated to make sure that enough fish are allowed to reach the spawning grounds to maintain the runs. In the absence of fishing, we expect the stock to fluctuate at relatively high abundance. When freshwater and ocean conditions allow the sockeye to grow rapidly and survive, more fish return from the sea than when conditions are poor. If there were no harvest, on average, for each fish that spawns, one will return. If more than one fish returns for each fish that spawns, the population increases. If less than one returns for each spawner, the population will decrease.

Only one fish per spawner returns, you say, that is all? Alas, a fish's life is fraught with danger from egg to spawner, and even though a female will leave about 2,000 eggs in the gravel nest and the male will fertilize them, if 4 or 5 years later one of the 2,000 returns, that is good odds.

The key to sustainably harvesting Bristol Bay fish is to know that all other things being equal, when the number of spawners is low, the lifetime survival rate increases, that is, there might be five or ten fish returning for every fish that spawned. The reason is location. When there are few spawners, each one can find the perfect gravel with the best freshwater stream flow that is ideal for their eggs. There will also be more food for each juvenile salmon growing in the lakes, and they will grow faster. Larger juveniles have a better chance of surviving in the ocean. When there is a surfeit of spawners, there is fierce competition for habitat; the nests of previous occupants are destroyed; and their eggs dispersed and eaten by gulls and char. In years of super-runs, like 2017, good real estate runs out quickly, and spawners are forced into marginal habitat, and this is followed by limited food for juveniles in the lakes. The result is lower lifetime survival for all. As we saw in the estimates of abundance in Lake Nerka from 1750 to 1900, before industrial fishing began, the number of fish spawning in the lake fluctuated around one million fish. One million spawners produced, on average, about one million returns.

Now It Is Time to Do the Numbers

Thanks to the long-term studies by the Alaska Department of Fish and Game, we have many numbers, particularly for the Wood River system in Bristol Bay.

The Wood River system is vast. It contains many rivers and lakes, forests and tundra, bogs and meadows, mosquitos and black flies, high mountains, small streams, and immense amounts of glacial gravel. It also contains the two field camps where I have so far spent twenty-five summers.

But onward to numbers, data, and a graph about the Wood River sockeye (Figure 2.1). Each point represents a year of spawning. The x-axis shows the number of fish that were counted coming into the lakes each year from 1963 to 2005. In the lowest year, 1973, only 330,000 fish reached the spawning grounds. The total salmon run that year was so small that almost no harvest was allowed. On the other hand, 1980 was a very good year, and, despite a significant harvest, almost 3 million fish forged ahead to spawning bliss.

The y-axis shows the ratio of adult offspring to the number of spawners. The adult offspring is the number of salmon who made it back into the Wood River after their time at sea, either to be caught or to escape to spawn. These salmon spend 1 year in the gravel as eggs, 1 year in the lakes, and 2 or 3 years at sea, returning from the ocean

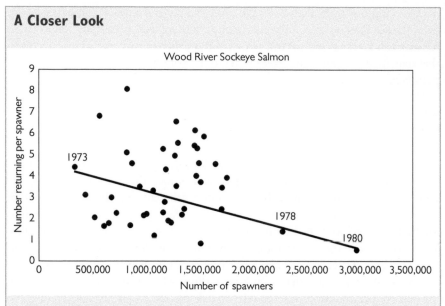

Figure 2.1 The number of adults returning for each fish that spawned during the Wood River sockeye salmon spawning years 1963–2005.

Each point represents the number of fish that spawned in a calendar year (the x-axis) and the number of fish that returned from those eggs (y-axis). The solid line is a possible relationship between the number of spawners and the average return per spawner.

as 4 or 5 year olds. The 1980 spawning cohort of 2.9 million fish produced 1.5 million adult offspring spread over the summers of 1984 and 1985. There had been just too much competition early in their lives, and they managed to produce only about 0.5 offspring per spawner. Another year with many spawners was 1978, when 2.2 million spawners produced 3.1 million adults. The solid line shows what the average number of returning fish per spawner might be like.

Another way to look at these results is the Table 2.1. It shows the return per individual spawner predicted by the line in the graph above for rising numbers of spawners. It also shows the number of fish that will return and, based on the returns, how big the harvest should be to ensure the return of the same number of spawners. Using these data, we would expect that, without any fishing at all, perhaps 2 to 3 million sockeye would return to the Wood River and each returner would have one offspring. We saw that in 1978 with a little over 2 million spawners 3 million fish came back, whereas in 1980 with 2.9 million spawners the return was only 1.3 million. So somewhere between 2 and 3 million spawners is the number that will produce one recruit per spawner, that is the number of returning fish will be the same as the number of spawners had been.

However, if we allow only one million fish to spawn and catch the rest, each fish would have more than one offspring, and on average 3.286 million salmon

Table 2.1 *The Return per Spawner, Average Return, and Potential Harvest of Wood River Sockeye Salmon, Determined Using the Data in Figure 2.1*

Spawners	Return per spawner	Return	Harvest
500,000	3.97	1,984,351	1,484,351
1,000,000	3.29	3,286,514	2,286,514
1,500,000	2.60	3,906,488	2,406,488
2,000,000	1.92	3,844,273	1,844,273
2,500,000	1.24	3,099,869	599,869
3,000,000	0.56	1,673,277	0

would return. That kind of return would allow us to catch an average of 2.286 million fish. If we consistently harvest all of the fish above one million, the sockeye stock would be sustained at about one million spawners and a harvest of 2.286 million fish.

For the sake of argument, what happens if we allow only 500,000 sockeye to spawn, and they had an average rate of 3.97 offspring each? The stock would then be sustained at about 500,000, with an average annual return of 1.984 million fish and an average annual catch of 1.484 million.

So what should the manager do? In Alaska the overriding goal of fisheries management is maximizing long-term catch, and the managers are tasked with letting the number of fish escape to spawn that will produce the highest average catch. Looking Table 2.1, we see that this would be 1.5 million fish.

Overfishing is a word that is used frequently but poorly understood. Simply stated, *overfishing* is harvesting harder, or allowing fewer fish to spawn, than would maximize the objective—in this case, long-term average catch. If fewer fish are allowed to spawn than would maximize long-term catch, it is called overfishing.

Managing an escapement target of either 500,000 fish or 1,500,000 fish is sustainable because both would maintain the spawning stock at the chosen level, but at 500,000, the catch is lower. Consequently, if we managed with an escapement goal of 500,000 fish, we would be overfishing, because, even though sustainable, we are not leaving enough fish to spawn in order to get the biggest possible annual catch in future years.

We must distinguish between sustainability, extinction, and overfishing. Because we can sustainably overfish, overfishing does not mean extinction. Through overfishing, we can drive a fish stock to low numbers but still be sustainable by the Brundtland definition, because year after year, that stock will produce similar amounts of fish for harvest. However, future generations might decide to allow more spawners, maximize yield, and stop the overfishing. In the Wood River system, for example, anything below 1.5 million would be overfishing, even though it would be perfectly sustainable.

From the Fish's Point of View, What Is the Good Life?

From a biological point of view, a good or successful life depends on how many of your offspring survive to breed themselves. We have already seen that securing good spawning gravel and lots of food in the lakes makes for higher lifetime offspring survival for the Wood River salmon. The technical term is *density dependence*, i.e., your success depends on the density of your species. More plainly put, the chance of survival and/or the number of offspring produced tends to be higher when there are fewer members of the species to compete with, having fewer competitors leads to a more successful life. Competition matters a lot in all stages of a fish's life. First in the survival from spawning to life in the lake (fewer competitors = better spawning sites and higher chances of the eggs hatching); then how big a juvenile fish in the lake can grow (fewer competitors = more food); and finally, how well that fish survives in the ocean before coming back to spawn. We know much less about the ocean life of salmon, but competition certainly plays a role there. When there are large numbers of fish of a given stock like Wood River sockeye, their average size when they return as adults is lower. There is good evidence that the overall number of salmon in the North Pacific also has an influence on both survival and growth, which is due presumably to competition for food. When there are fewer competitors at all life stages, we get more and bigger fish who beget more than one offspring, possibly three to four in the case of Wood River sockeye.

When This Is the Case, How Many Fish Can a Fisherman Fish to Have Plenty of Fish Again Next Year?

That is, how do we sustainably harvest fish? Almost all fish species for which we have long-term data show reduced survival and/or growth at high densities. If we harvest a greater part of a mature population of fish, there will be fewer fish left, and the lack of competition will allow those fish to produce more and larger offspring. The fisherman's and the manager's conundrum is to find the sweet spot between high annual harvest and survival of enough fish over the long term.

Maximum Sustainable Yield: Magic or Curse?

One of the most important and controversial topics in fisheries sustainability is *maximum sustainable yield* (MSY). This calls for a graph (Figure 2.2) that might be called "Go Find the Sweet Spot."

The fish population at the sweet spot is called *Biomass at MSY* (BMSY). It is a bit like Goldilocks' bed: not too big, not too small, but just right. It is fishing at the level that produces the maximum number of fish caught sustainably over the long term.

Anything to the right of the peak of the thick line is overfishing because you could catch more fish over the long run if you took fewer fish at the moment.

A Closer Look

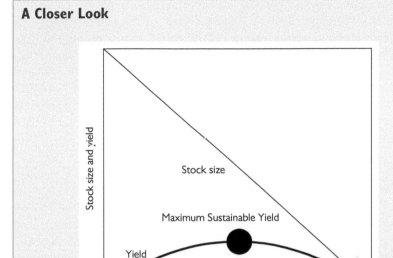

Figure 2.2 The relationship between population size and average harvest.

The large dot indicates the maximum sustainable yield. The thin line is the number of fish in the population. The thick line represents the fish that are being caught. True to our previous argument, there will be fewer fish in the water the harder we harvest, and yet the catch keeps increasing up to some point. If we continue to fish harder, eventually both stock and harvest go down. We missed the sweet spot.

MSY was born in the 1930s, in the early days of analyzing fisheries population dynamics, and it is still controversial because the purpose of MSY is to get the best possible long-term catch, without any guarantee of economic profitability, community stability, or environmental protection. MSY seemed outdated by the 1960s, and Peter Larkin of the University of British Columbia wrote *An Epitaph for the Concept of Maximum Sustainable Yield.*[3] However, as nations were negotiating the Law of the Sea in the 1970s, MSY was the one target for sustainable fisheries management that could generally be quantified. Despite having been buried by Larkin, it was written into the Law of the Sea and adopted for want of anything better by many national legislations as the goal for their fisheries management.

John Gulland, who was arguably the most important fisheries scientist of the 1960s and 70s, once sent me a tongue-in-cheek definition: *MSY—A quantity that has been shown by biologists not to exist, and by economists to be misleading if it did exist. In short, the key to modern fisheries management.*

Another reason why MSY vexes people is that fish numbers are influenced by many factors besides fishing pressure. Fish have good years and bad years. There is no magic number of fish that can be safely harvested year in and year out with the guarantee that the fish population will not decrease—unless that number is really small. Since fish populations fluctuate depending on predation, competition, ocean conditions, etc., you do not get to the MSY sweet spot by harvesting the same number of fish year after year. MSY is not graven in stone. When conditions are good, you get to MSY by fishing more and harder. When conditions are bad, you have to grit your teeth and reduce the catch. Getting to MSY is a management process that must constantly keep track of any changes in both population size and productivity and make the appropriate changes to catch levels.

Environmental Changes and Regime Shifts

There is no shortage of challenges in managing fish populations sustainably, and one of the more baffling is that many populations oscillate wildly in productivity. Wood River sockeye, as we have seen, might produce from one to seven offspring per spawner—that is a lot of variability. And it gets worse. Some populations have very long stretches of good years, and equally long stretches of bad ones, called *regime shifts*. Take the California sardines. We know what they have been up to for the last 1,000 years because someone took sediment cores from the bottom of the ocean going down that far and counted the number of sardine scales in them. For decades, there was not a single scale, and then for another few decades, there were nothing but sardine scales—and that was long before there was any sort of large-scale fishing pressure. Sure enough, sardine populations in other locations showed us that this pattern of long periods of good and bad conditions over centuries is not unusual.

Well, what about MSY with this kind of fluctuation? Trying to find the level of fishing pressure that maximizes long-term yield over long periods of great abundance and long periods of none whatsoever is extremely challenging. Might as well admit that there is an MSY during the good times but there is no level of harvesting that will produce MSY during the bad times. Using MSY as a management tool for many fisheries has limitations because though sardines may be the extreme, roughly half of the fish populations that have been analyzed go through these sorts of regime shifts.

Multiple Stable States

Knowing that sardines and other stocks move in and out of periods of abundance, it became clear that ecosystems can be stable in either of those states. We call that *multiple stable states*.

For instance, imagine a population at high abundance that has sufficient food and predators that take a certain number of individuals per year. If nothing changes, that is not a worry. But if a new force like fishing drives the population down too much, suddenly the number the predators are taking is a big chunk of the population. Now we are in a new steady state, where the prey are at low abundance, limited by steady predation, which in turn means there is little or no sustainable yield. It is possible that, even if all harvest is stopped, this limited steady state will last forever, i.e., the population will not be able to grow and flip back to a state of high abundance. In the worst case, it may be driven extinct. This is exactly the opposite of the density dependence we saw in Wood River sockeye, where survival increased as density went down. There is not a lot of evidence for survival rates declining at very low densities, but there are certainly reasons to think it may happen.

Sea otters, sea urchins, and kelp forests have been jumping around between states for quite a while. Originally, otters munched on sea urchins who grazed on the vast kelp forests and all lived in harmony in the Pacific Northwest. Enter the Russians, with guns and the craze for winter coats and hats made of the exquisitely beautiful and equally exquisitely warm otter pelts, and the otters nearly disappeared in the nineteenth century. The uneaten sea urchins became legions marching along the bottom, clear-cutting the kelp forest. A new state of few otters, lots of urchins, and little kelp arose. When otter hunting became unprofitable and stopped, the otters had a chance to recover. They did recover in some places, but not in others, like Southern California. Consequently, there are now patches of different kinds of otter-urchin-kelp stable states.

The moment when too few otters can no longer control a sea urchin explosion is called the *tipping point*. Identifying if and where tipping points might happen is one of the bigger challenges in sustainable management of not only fish stocks but ecosystems in general.

Expanding to Ecosystems

So far we have looked only at the sustainability of a single target species. But fish populations do not live in isolation, and when you fish for one, you will catch or affect others. For example, nontarget species will be caught (called bycatch), some fishing methods like trawling can damage sensitive habitats, and removing fish from the population affects both predators and prey of the target species. Let us assume we can actually estimate how we would affect other elements of the ecosystem with the management strategies we could use for the target species, and then we need to go back to our definition(s) of *sustainability*. If the "inclusive measures of well-being for people" are our objective, we will need to evaluate how different levels of harvesting our target species will affect that well-being through such indirect pathways, and whether our decisions can be sustained over time. If, for instance, the best possible harvest of a target species would greatly reduce quality and abundance of a sensitive habitat, but that habitat is important to the well-being of people, then the best possible harvest is not sustainable. The same is true for impacts on bycatch and the predators of our target species. First we would need to clearly define *well-being of people* before we can determine what is sustainable.

But with the Brundtland definition, "meets the needs of the present without compromising the ability of future generations to meet their own needs," we can look at the matter differently. Depleting the abundance of a highly valued, sensitive habitat will definitely reduce well-being. But if that habitat can recover if fishing were stopped, then, by the Brundtland definition, our decision to deplete would still be sustainable because future generations could choose to stop fishing, and the sensitive habitat will recover. Our depletion of it would not prevent future generations from enjoying it, although I think the rub here is in the definition of future. If it takes centuries for the habitat to return to a productive state, and then some future generations would not be able to meet their needs, we are on slippery grounds with sustainability.

Sustainability Certification and Labeling: Red Fish, Green Fish?

So How Does Someone Who Is Not a Fisheries Scientist Know Which Fish Populations Are Sustainable and Which Are Not?

By the time fisheries sustainability entered the consciousness of the informed and somewhat affluent public, quite a few NGOs were ready to give advice and provide certification, telling us what we should and should not eat. All kinds of pocket guides to take to the fish counter turned up with green = good fish; yellow = okay fish; and red = do not eat under any circumstances. Fishmongers started to label their daily catch accordingly. But soon enough confusion set in because one certifying NGO's green fish was another one's red fish.

One had to conclude that the green-yellow-red classification was subjective. Even though there is certainly plenty of room for scientific analysis, each NGO set its own standards of what it considered acceptable environmental impacts or acceptable levels of fishing pressure. How a fish was classified was the opinion of the maker of the guides and not always science based. So how would the largest commercial fishery in the United States, Alaska pollock, be judged? The Marine Stewardship Council certified it as well managed: pollock on the green list. The Monterey Bay Aquarium's edict was "a good alternative": pollock on the yellow list. Greenpeace deemed it unacceptable, and our pollock was shamed onto the red list.

Throughout the world, there is a lot of conflicting advice like that about the same fish. In the United States, the certifiers with the most influence are the Marine Stewardship Council (MSC) and "Seafood Watch" of the Monterey Bay Aquarium. They are so influential that the retailers in North America and Europe pledged to sell only wild-capture seafood that meets the standards of either one.

Because of their transparency and their wide acceptance by retailers, we will look at them more closely. Their methods of classification are well documented, and both agencies' criteria are strictly concerned with marine ecosystems. But they do differ dramatically: MSC is about process, whereas Seafood Watch is about state.

MSC scoring criteria ask questions about how a fishery is managed. The criterion for bycatch reads, "There is a strategy in place for management bycatch that is designed to ensure the fishery does not pose a risk of serious or irreversible harm to bycatch populations." This is evaluating the process of fisheries management.

The equivalent for Seafood Watch, on the other hand, is "Stock abundance and size structure of all main bycatch species/stocks is maintained at a level that does not impair recruitment or productivity." So Seafood Watch is taking a snapshot of abundance of nontarget species, while MSC is asking if the system will, over time, be sustainable. I think MSC is definitely a better way to judge sustainability.

However, the impacts of fishing go beyond the simple removal of fish from the sea. Neither one of these certification programs has traditionally been the slightest bit concerned with social or economic factors, nor the interaction of fish products with terrestrial ecosystems. In fact, no fish certification programs consider all the factors that could affect sustainability, whether it is defined as the ability of future generations to meet their own needs or as all-inclusive measures of human well-being. In the last year or two, though, there has been some change as human slavery began to be considered in many certification and labeling programs.

Beyond Biological Sustainability

Fishing is not just about fish. Fishing is about feeding people and providing employment. The current overwhelming emphasis of managing fishing not for social or economic benefits but for biological sustainability prevents us from addressing the needs of both humans and the ecosystems they rely on.

On an international level, in 2015, the United Nations published its Sustainable Development Goals (SDGs) that were agreed on by 197 nations, and include seventeen general goals and 169 specific targets. Many of the terrestrial goals are social and economic, such as:

Goal 1: No Poverty
Goal 2: No Hunger
Goal 5: Gender Equality

But once we get to fisheries and oceans, it is biology only. Under "Goal 14: Life Below Water," we find SDG 14.2, "by 2020, sustainably manage and protect marine and coastal ecosystems to avoid significant adverse impacts, including by strengthening their resilience and take action for their restoration, to achieve healthy and productive oceans."

SDG 14 is all about the biological status of ecosystems. Two specific elements of sustainability are (1) the share of coastal and marine areas that are protected, and (2) the percentage of fish tonnage landed within MSY. Once again, social and economic elements of fisheries sustainability are not even on the radar.

This is truly baffling since fisheries exist to produce benefits to people. Any manager, be it for large industrial fisheries in rich countries or small-scale fisheries in poor countries, will consider social and economic factors in the sustainability equation. Social scientists, on the other hand, often lament the sometimes-negative social and economic impacts of fisheries regulations designed around environmental sustainability.

Serge Garcia is a French scientist whose early professional postings were in West Africa, where French agencies still maintain a strong presence. He told me about fisheries near big cities in Senegal, where a rich man may own a large net, called a beach seine, used to encircle fish near the beach. A small boat drags the net offshore, and then dozens, and sometimes hundreds, of people on the beach pull the net in by a long rope because if you can get a hand on the rope, you get a fish. Many a biologist has tried to get rid of the beach seines because they catch too many small fish—and that is a biologically inefficient way of fishing. Garcia, however, sees it differently. Fishing like that is vital to a *de facto* social support system. People without a job can spend the day on the beach, and if a seine needs to be hauled and they get a hand on the rope, there will be a fish to eat. Without the nets, would they starve? Garcia wonders. What is the appropriate balance between ecological or economic maximization and social impact?

What about the future and the social sustainability of existing fishing communities, be they in Iceland or Norway, Samoa or Indonesia? The Palau fishermen with larger diesel powered boats encroach on the traditional fishing areas of villages with small boats. In Indonesia and Brazil, industrial fleets from big cities take the fish that small villages used to catch. In West Africa, factory ships from Europe and Asia deplete the resources traditionally caught by small-scale fleets. In Europe and North America, small-scale fishermen are being replaced by industrial fishing companies. All over the world, the economic and social sustainability of fishing communities is under threat.

The reasons are complex but unsurprisingly have to do with money translated into political power and economic efficiency. Large industrial boats are economically more efficient than small ones are. Large-boat industrial fishermen are richer, better organized, and politically more powerful than small communities are. There is nothing unusual about this process. Small farms have been replaced by large farms; small retail stores by larger ones. Maintaining small-scale fishing is a choice that some societies may make, but others may not.

Only recently have we started to consider some of the complex social aspects of fishing. In 2014, the practice of slavery in fisheries and fish processing plants, where workers were often held against their will, was exposed by popular media. In 2018, Seafood Watch alerted consumers to the probability of human rights abuses in certain fisheries. That was the first time any aspects of social sustainability in fisheries entered the public discourse.

When academics and the public talk about sustainable fisheries, they talk about marine ecosystems. When NGOs rate and certify seafood, they talk about marine ecosystems. Only recently has discussion begun about energy use and greenhouse gases in the context of sustainability in fisheries.

Fisheries are complex human enterprises. Sustainability in fisheries needs to reflect that, to expand beyond aquatic ecosystems to encompass social and economic measures. Sustainability is determined primarily by the process of management. If abundance of stocks can be measured, and fishing pressure can be adjusted in response to changes in abundance, then the fishery is likely biologically sustainable. Sustainability involves producing goods and services that we want, with the constraint that we maintain the productivity of the natural ecosystems on which we depend, and it should be measured primarily by the continued production of those goods and services. However, in addition to managing the biological resources, there is a wide range of decisions to be made about who gets to fish, what kind of gear is used, and how the catch shall be managed that will likely have much more impact on the social and economic sustainability.

• •

FURTHER READING

The Brundtland report on sustainability. Brundtland G. 1987. Our common future: Report of the 1987 World Commission on Environment and Development. United Nations, Oslo, 59.

A book looking at all aspects of sustainability. Matson P, Clark WC, and Andersson K. 2016. Pursuing sustainability: A guide to the science and practice. Princeton University Press.

A classic paper describing the end of maximum sustained yield (MSY) as an objective. Larkin PA. 1977. An epitaph for the concept of maximum sustained yield. Trans Am Fish Soc 106: 1–11.

A tongue-in-cheek look at why MSY came back from the dead. Smith ADM, Reynolds JD, Mace GM, et al. (eds). 2001. The gospel of maximum sustainable yield in fisheries management: Birth, crucifixion and reincarnation. In: Conservation of exploited species. Cambridge: Cambridge University Press: 41–66.

A more in-depth look at fisheries sustainability. Hilborn R, Fulton EA, Green BS, et al. 2015. When is a fishery sustainable? Can J Fish Aquat Sci 72: 1433–41.

CHAPTER 3

How Fisheries Are Managed

Traditional Community-Based Management

The people of the Pacific Islands of Polynesia, Micronesia, and Melanesia depend on the sea for their survival. Before Western contact, most of their protein came from fisheries; even today, with modern trade, many of these countries get over 50 percent of their protein from their own fisheries. They managed their fisheries sustainably for thousands of years; those who did not were unlikely to survive.

The seminal work on community-based management was *Words of the Lagoon* by Bob Johannes,[1] published in 1981, which described traditional fisheries management in Palau. Johannes spent many years working in a number of Pacific Islands studying how the islanders managed their fisheries and found that there were common threads throughout the area.

> The most widespread single marine conservation measure employed in Oceania, and the most important, was reef and lagoon tenure. The system was simple: The right to fish in a particular area was controlled by a clan, chief, or family, who thus regulated the exploitation of their own marine resources. Fishing rights were maintained from the beach to the seaward edge of the outer reefs. It was in the best interest of those who controlled a given area to harvest in moderation. By doing so they could maintain high sustained yields, all the benefits of which would accrue directly to them.[2]

Local chiefs or heads of clans, who had tenure and exclusive access to their fishing grounds, used a variety of tools to regulate fishing pressure. They could close areas, close seasons, close time periods, allow a portion of potential catch to escape, declare size limits, and restrict the number of traps.

Johannes concluded that over the range of Pacific Islands, and prior to the arrival of Europeans, these methods were generally effective, and they maintained the supply of seafood. He then described the destruction of these traditional management systems by colonialism that replaced them with Western-style government management that largely failed. Toward the end of his career, Johannes documented a partial revival of many of the aspects of traditional management, and, as we shall

Ocean Recovery: a sustainable future for global fisheries? Ray Hilborn and Ulrike Hilborn, Oxford University Press (2019). © Ray Hilborn and Ulrike Hilborn 2019. DOI: 10.1093/oso/9780198839767.001.0001

see, the key concept of exclusive tenure by local communities is now widely accepted as salient to sustaining small-scale fisheries around the world.

The Western Model and Freedom of the Seas

Western fisheries management has followed a very different course, initially based upon the concept of "freedom of the seas" that allowed free access and unrestricted fishing beyond the territorial seas of individual nations. *Territorial seas* were commonly defined as 3 miles from the coast, based on the seventeenth-century distance a cannon could fire, and was upgraded to 12 miles as cannons became more powerful. International waters remained unrestricted, and even within the territorial seas of most Western nations, there were no restrictions on fishing.

The Prime Example of Unrestricted Fishing Is Whaling

Oceanic whaling began sometime in the eleventh or twelfth centuries when the Basques hunted right whales in the Bay of Biscay.[3] Right whales were so named because they moved slowly and did not sink when killed, and thus were the "right" whales to attack. By the thirteenth century, the Basques had depleted the local stocks close to their fishing ports and had to move farther from home in search of new stocks. Technological innovations allowed them to catch and process whales at sea by the fifteenth century and thereby inaugurated modern whaling. Holland and England adopted the Basque fishing techniques, and by the seventeenth century, right whales were scarce throughout the North Atlantic. Whalers began to move around the world, sequentially depleting the whale resources until each hunting ground became unprofitable, forcing them to find the next. By the end of the nineteenth century, traditional whaling was in crisis. Stocks were depleted, petroleum began to replace whale oil, World War I put an end to corsets that needed baleen stays, and prices tanked. Human ingenuity intervened, and the technical innovations of steam-powered catcher boats, explosive harpoons, and air bladder inflation of large whales that would normally sink led to a boom in Antarctic whaling that brought the larger Antarctic whales close to extinction.

There were no restrictions on whaling, just as there were few if any restrictions on marine fishing throughout the Western world. Inevitably, accessible and valuable fisheries were being depleted, and soon it was obvious that regulations were needed and put in place. The history of fisheries management in the territorial seas of Europe and North America, as well as former British colonies of Australia, New Zealand, and South Africa, has followed a similar sequence. As long as fishing gear was inefficient, sailing ships using hook and line could not exploit a significant fraction of the fish stocks. Thomas Huxley, in 1884, reflected a common view of the vastness of marine fish resources and scorned regulations in view of the forever-abundant great sea fisheries.

I believe, then, that the cod fishery, the herring fishery, the pilchard fishery, the mackerel fishery, and probably all the great sea fisheries, are inexhaustible; that is to say, that nothing we do seriously affects the number of the fish. And any attempt to regulate these fisheries seems consequently, from the nature of the case, to be useless.[4]

However, in the next line of his address, Huxley says, "There are other sea fisheries, however, of which this cannot be said." He recognized that oyster beds can be exhausted, as can salmon fisheries. So while he is often maligned as having been ever so wrong, in fact, he was largely correct because, given the technology of his time, many of the major marine fisheries were unlikely to be depleted. This changed rapidly with the advent of steam-powered vessels, modern fishing nets, and canning and refrigeration that subsequently created a much greater market for fish.

Consequently, not until fish were sufficiently depleted to threaten the economic viability of a certain fishery, did it become obvious that too many fish had been taken and that it was time for regulation.

One of the earliest documented cases was the Norwegian cod fishery in Lofoten that exploited the Northeast Arctic cod stock, the largest cod stock in the world. The traditional fishing gear was jigging, holding a line with a baited hook held down with a rock. But the 1700s brought new gill nets that catch fish by their gills when they try to swim out of the net. Another new method were longlines, an especially long fishing line with many baited hooks that is left in the water overnight. Soon worried fishermen banded together to petition the King to ban these newfangled methods, and in 1744, the King directed the governor of the region to allow only the traditional hand lines. This kind of restriction of efficient gear became the dominant initial form of fisheries management throughout the Western world.

Unusually for European fisheries, the Lofoten fishery also experimented with the tenure system described by Johannes for the Pacific, a system that is now often called TURFs or Territorial User Rights to Fish.

> In 1816, after the restoration of the Norwegian state, albeit in a union with Sweden, a law on the Lofoten fishery was enacted. This law went a long way towards establishing what lately has come to be known as territorial use rights. The fishing banks were divided into areas belonging to the nearest fishing base on land and further subdivided into fields where the boats were allowed to fish. The allocation of the fishing fields was in the hands of local governing committees, usually headed by the owner of the onshore facilities which the fishermen had to rent for accommodation and for drying the fish.[5]

Fur seals, a high-value resource in the eighteenth century, were another early target for regulations. When the females hauled out onto the beaches with their pups, they were extremely vulnerable and easy to hunt. To satisfy the Chinese market, fur seal colonies throughout the world were wiped out almost as soon as they were discovered. The exception was a colony that survived on the Pribilof Islands in the Bering Sea, where hunting females had been banned by the Russian administration in 1834. In 1867, Russia sold Alaska to the United States, Congress passed legislation to protect these breeding colonies in 1868, and in 1874, subsequent legislation authorized setting of quotas on how many could be taken and determined the seasons when fur seals could be hunted.

Modern Western Fisheries Management

That fisheries management is a permanent work in progress is neatly illustrated by Australia, which followed closely a pattern of events we find wherever a central government has taken charge of its fisheries. It begins with (1) the recognition that a stock is depleted and that it is time to curtail harvesting, and then it moves on to (2) restrictions, who can use what fishing gear, protecting breeding or rearing sites, and setting minimum-size limits. If things do not get any better: (3) governments move on to limiting the number of vessels allowed to fish and sometimes "buy-back" vessels or fishing permits, followed by (4) total-catch quotas and allocating portions of that catch to different gears or areas, and (5) allocation of the catch to specific groups, individuals, or companies.

A practical example is the prawn fisheries of West Australia. Nursery areas were closed as early as 1963, and in the same year, the number of boats allowed to fish was limited. The size of nets was regulated in 1976; closed seasons arrived in 1982; and in 1985, the fleet was reduced by buy-back of vessels. Much like the Bristol Bay sockeye fishery discussed in Chapter 1, the catch of these prawn fisheries is now regulated by time and area closures to ensure that a sufficient number of prawns survive to reproduce.

In multination fisheries, the case is altered. The first step must be allocation of fishing rights among the nations involved. Without this allocation, there can be no agreement on management.

The Tasmanian abalone fishery has a much more involved regulatory structure. It did not really get going until the 1960s, when hookah diving gear was invented, which allowed the divers to stay under water longer to pluck abalone off the bottom one by one to send to the high-end markets in Asia. It became obvious quite early that the number of boats and the size of abalone had to be restricted. Fishing licenses were first issued in 1965, and minimum-size limits in 1967. In 1969, a freeze was placed on new licenses with an annual fee of $100, and all divers had to prove they earned most of their income from abalone fishing. If a diver left the fishery, a new license would be issued to someone on the waiting list. In 1974, licenses became transferable, that is, they could be sold, and the price for a license immediately rose rapidly to $100,000. Because in 1974, $100,000 was still an awful lot of money, the new divers had to work harder to pay off their loans, and total fishing effort rose rapidly. To counteract these dangerously high catches, Tasmania introduced Individual Transferable Quotas (ITQs). Now each diver could catch no more than an allocated, limited number of tons of abalone but could catch them when and where the diver wished.

However, human nature being what it is, as soon as ITQs became very valuable, control of quotas moved from individual fishermen to processors or other moneyed interests. This is commonly seen in ITQ fisheries, and pros and cons of ITQs are discussed in Chapter 4.

The Tasmanian abalone fishery illustrates the shift from input to output controls that has happened in many managed fisheries. *Input controls* are the number of vessels, the kind of gear, licenses, area restrictions, and seasonal openings. *Output control* is the number of fish caught.

To control output, managers set a number, the total allowable catch (TAC), at the opening of a fishery, and when that number is reached, the fishery closes for the year. In an "Olympic" management system, all of the licensed vessels compete with each other to catch the largest share before the season closes. A different way to control output, used in the Tasmanian abalone fishery, is to divide the TAC among individuals or groups.

Almost all national fisheries agencies use some variation of simple input controls, including limitations on gear, a limited number of licenses, and time and areas closures. The extent to which these regulations are effectively enforced differs greatly from country to country.

Output controls are fairly common in most United States, New Zealand, and European Atlantic fisheries. Some, particularly Iceland and New Zealand, and many US federally managed fisheries have used individual or group allocation, whereas within the European Union, it is common to allocate catch to groups known as "producer organizations."

The Bristol Bay sockeye fishery of Chapter 1 limits the number of fishing permits, the size of boats, and the size of nets, and it regulates openings and closures for each area. Robert Johannes points out that Pacific Islanders would be extremely baffled by any attempts to reduce the efficiency of fishing—input controls are unknown there.

There is a strong dichotomy between Western-style, top-down fisheries management that relies on a central regulatory agency, usually a government fisheries department, and true community-based management, where exclusive control is granted to the community, either legally or by custom, and regulations are set and enforced by the members of the community.

Modern Community-Based Management: What Works and What Does Not

Mekong River

Freshwater fisheries rarely make headlines, but they feed millions of people around the world and often rival marine fisheries in catch and employment. For the time being, we will relegate recreational fisheries to its own chapter.

The Mekong River runs 4,300 kilometers from the Tibetan Plateau through China, Myanmar, Laos, Thailand, Cambodia, and Vietnam, and it gives food security to millions of people in its catchment. While data are not very accurate, an estimated 2.5 million tons of fish and other aquatic species are landed each year, and about 40 million people are engaged in wild-capture fisheries. Throughout most of the Mekong, these fisheries are essentially unregulated, but there are recent attempts at community-based management.

In the Khong District, Campasak Province of Southern Laos, a community-based system was set up in sixty-three villages between 1993 and 1999. Regional authorities handed over management of the fisheries near their villages to the communities who instituted several rules: (1) deep-water no-take zones to allow a refuge for fish in the dry season; (2) a ban on blocking streams with fish traps; (3) a ban of specific fishing methods; (4) a ban on catching juveniles of certain species; (5) protection of certain species during spawning or juvenile rearing seasons; and (6) certain forms of habitat protection, all based on specific knowledge held by local people on the biology and ecology of the fish, also called traditional ecological knowledge.

Japanese Cooperatives

In Japan, cooperatives have had legal management authority for quite a long time. Before 1900, there was a jumble of local and national regulations and structures, largely ineffectual and conflict ridden. The Fisheries Law of 1901 created exclusive fishing rights for community cooperatives. Subsequent legislation set up what are TURFs, that center around formally designated Fishing Cooperative Associations (FCA) in each community. Each FCA owns its coastal waters with the exception of ports and industrial zones, and sets local fishing regulations for their members that include gear restrictions and time and area closures. The national and regional governments provide some support for these FCAs through science advice and some subsidies. Recently, though, the cooperatives seem to be in serious trouble. Two-thirds show a financial loss, and cooperatives are merging. There were more than 2,000 in 1989 and fewer than 1,000 in 2016.

Chilean Artisanal Fisheries Management

The *loco*, a large marine snail, is a delicacy in its native Chile and in Asian markets, where it competes with abalone; 10,000 licensed divers collect it along much of the 4,270 kilometer-long coast of Chile, where it is the major source of income to hundreds of small fishing communities.

During the 1980s, the loco fishery was worth roughly $US 64 million per year. In theory, it was regulated with a minimum-size limit and limited entry. In practice, there was no enforcement; stocks were depleted; and the central government closed the fishery in 1989.

No hope for loco? Not exactly, because Juan Carlos Castilla and Carlos Moreno came to the rescue. They were both marine ecologists at Chilean universities. Castilla did much of his work at the university's research lab 2 hours west of Santiago at a small village called Las Cruces. He had convinced four local fishing cooperatives, known as *Caletas*, to keep the area in front of the marine lab closed to loco fishing. He also demonstrated that closed areas could restock the depleted ones, and that the value of the loco could not only be increased, but the fishery could be sustained with good management. Largely as a result of his work and that of Moreno, the Chilean government passed a law in 1991 that allows Caletas exclusive access to management and exploitation areas (MEAs), if a range of conditions about monitoring, management plans and enforcement were met.

In 1999, I joined a team that reviewed the impacts of the 1991 law. A relatively small number of Caletas had enrolled in the MEA program by then, but studies by Castilla and Moreno showed that for those who had joined, the size and abundance of loco had increased, and the fishery was becoming more profitable. Particularly in the Caleta El Quisco, just north of Las Cruces, which worked closely with Castilla, the catch per hour of diving had soared dramatically compared to the catch per hour outside. Based on this success, we recommended that the government do all it could to expand participation in the program.

By 2007, 732 areas had been designated for entry into the program, and 237 already had a management plan in place. However, the outcomes are not all good. Most of the coastline remains outside management areas. All licensed divers have open access there and, as might be expected, loco stocks there are again depleted. Inside the managed areas, success is also a mixed bag. Caletas with big areas and a rich resource will support its members well, but many areas have failed, and even the Caleta El Quisco, along with many others, has withdrawn from the system.

One of the most successful examples of community-based management in Chile is in the Juan Fernández Islands, 670 kilometers off the Chilean coast. The islands are far enough from the mainland that the fishermen have exclusive access to the spiny lobster, their primary target, which they catch with traps. This Caleta has formal rules, including a legal size, closed season, and release of females with eggs. Diving for lobsters is banned. But there is also an informal form of individual tenure of fishing locations. Individual fishermen or families effectively own individual fishing spots, where they set traps, one trap per spot. Both use and transfer of spots are governed by informal but widely acceptable rules.

The Chilean experiment has taught us a lot about how useful TURFs are for small-scale fisheries. At present, the Caletas with sufficient area to make a living and the internal leadership and cohesiveness to maintain control are still doing well. However, the loco fishery has been officially closed for the rest of the coast, but, unofficially, everyone is poaching, taking undersized loco, and contributing to the depletion of the fishery.

So What Works in Community-Based Management?

Elinor Ostrom won the Nobel Prize for her work on community management of natural resources. While her work concentrated on forestry, it is equally relevant to fisheries, and her conclusions have been largely validated for fisheries. Successful community-based management depends on several factors.[6] First of these is exclusive access—if a community cannot control its fishing grounds, if others can come in and take fish outside its regulatory system, community-based management will fail. The next most important factor is leadership and cohesion within the community. Without strong leaders and a cohesive community, collective action is difficult to organize. Finally, the Chilean example clearly illustrates that there have to be enough resources to support the community within their management area. If the area under exclusive access is too small, the people of the community will also need to fish in outside areas, and the importance of the community fishery may not be significant enough to them to cause them to devote the energy necessary to manage their own.

The Process of Management: Choosing Regulations

How Does a Management Agency or Community Choose Various Options to Regulate from Those That Are Available?

The first step is to understand how we determine the trend and status of a fish stock or a community of species of fish, and the second step is how that information is translated into management actions such as size limits, area closures, time closures, gear restrictions, and catch limits.

Two kinds of biological knowledge are used to make the rules. In small-scale and community-based fisheries, this information comes from fishermen with traditional ecological knowledge. They know where fish spawn, where the juveniles are found, and where the mature fish are easily caught. In countries with centralized institutions, scientists study the biology and distribution of fish. Traditional knowledge or scientific findings are often used to determine a legal size limit and openings and closures of seasons and areas. In many fisheries, there is a total-catch limit—the fishing season is closed whenever the catch limit is reached.

Why Is Size Limit Important?

If we want to catch fish next year, there must be enough fish in the water that are mature enough to lay eggs this year. We can achieve this with lowering fishing pressure or using a no-take policy below a certain size, and that size should be larger than the size at reproductive age. Size limits work very well with invertebrates like lobsters, crabs, abalone, and sea urchins because undersized individuals can be released alive and grow large enough to reproduce.

Lobster and crab fisheries illustrate this very well. Baited pots are built with one-way entry doors that allow all sizes in, but they have escape doors that allow the very small ones to get out, and those that are undersized will be released. Many lobster and crab fisheries find that a size limit is sufficient to maintain healthy populations.

With abalone, sea urchins, and many snails that are collected by hand, a size limit will keep them reproducing as long as the divers recognize the correlation between size and making eggs. Jeremy Prince, an Australian biologist who works with small fishing communities in the Pacific, encourages divers to take a range of sizes of their catch, take out the gonads, and lay them alongside the animal. He can then show them that small animals often have no mature eggs, whereas large animals have lots of them, and the connection between size and fertility is made.

Counting What Is Invisible and Moves

It is really challenging to manage a fishery if we do not have a good idea of either how many fish there are out there or the trends in their abundance. With that knowledge, we can predict how many fish can be caught and how many need to stay in the ocean in order to keep populations healthy. Knowing how many fish there are is vital for management, however as the British scientist John Shepherd said, "Managing fisheries is hard: it's like managing a forest, in which the trees are invisible and keep moving around."[7]

How do we count fish? We do not—we estimate, extrapolating from a range of information collected in various ways. One vital piece of information for managers is how many fish are actually caught. Catch data are accumulated either by requiring fishing boats to keep logbooks or by sampling the landings of vessels when they return to port. From these data, managers estimate the total catch per boat or per day fished, which tells them how successful a fishery is. If there are changes, it may indicate a change in abundance.

Trends in abundance are estimated by either catch per boat or scientific surveys. Scientific surveys differ from using fishing fleet catch data because surveys cover the entire range of a fish stock, whereas fishermen tend to go where they think abundance is highest. The surveys give the managers the relative index of abundance that can be reliably tracked over the years. This is the method used in many industrial fisheries in much of the world.

Surveys that cover the whole range of the fish stock are vital because when fishermen stick to concentrations of fish, their catch per hour may stay high, even if the overall abundance of the fish is declining. Fishermen who have been onboard survey vessels often complain that the survey goes where there are no fish. However, covering the whole area used by the fish stock is essential for watching long-term trends. Surveys need to cover where fish actually are, as well as where they are not, in order for the index of abundance to be reliable.

Measuring Length and Dating Ear Bones

Another critical type of data is the size and age of the fish from the catches and surveys. Measuring size is easy. Determining age is more complicated. First, you have to cut the ear bones (*otoliths*) out of each fish's brain, and then the otoliths have to be ground down thin enough so you can see and count the growth rings that are much like those of a tree. The resulting age data are then used to track not only how old the fish are but also which years produced many offspring and which were reproductive failures. The otoliths of fish in temperate regions show changes in growth during summer and winter, which makes them easy to age. Tropical fish that have no winters are much more difficult to age, and not many attempt it. Aging of fish is usually done by laboratories run by national fisheries agencies.

These kinds of data are collected for perhaps 1,000 to 2,000 individual fish stocks around the world and are used in what is known as a "stock assessment," which estimates the trend in the abundance of a stock over time. Data from stock assessments have been assembled by an international team and are available for stocks representing a little over half the landings reported to FAO. These data are the basis of much of the analysis in Chapter 5 on the status of fish stocks.

What Is Working and What Is Not?

Nobody has yet systematically analyzed how fisheries around the world are managed and what really works to make fisheries sustainable biologically, socially, and economically. We know that there is a clear relationship between the intensity of management and the fishery's biological status,[8] but we also know that the social and economic performance of fisheries may be largely unrelated to the biological status. Depleted fisheries may still be profitable and provide benefits to the community, even though there could be more fish if the stock(s) were more abundant.

Western-style, top-down management works well for rebuilding and maintaining healthy stocks when there is science involved and the scientific advice on catch levels is followed and enforced (see Figure 3.1). Some countries, like Chile, are trying to go that route, but the difficulties are that this kind of management only works when there is a strong central government with a well-funded science and management agency that has secured the cooperation of those doing the fishing. European Atlantic fisheries are generally rebuilding under this approach. European fisheries

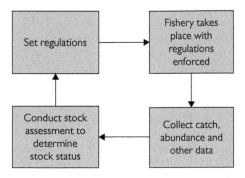

Figure 3.1 The cycle of fisheries management when central governments collect data and set and enforce regulations.

in the Mediterranean are almost all small scale and overexploited. Because there are up to twenty nations involved, unified management is clearly impossible.

For such small-scale fisheries, the top-down approach is neither possible nor desirable. The cost of managing thousands of small fisheries is too high, and it is widely accepted that various forms of community-based management are the only option. This is true even for countries with sophisticated management systems. In New Zealand, for instance, about 500 individual species area stocks have been defined, and an annual allowable catch is set for each one. Yet the only ones assessed and managed intensively are the thirty fisheries that constitute most of the total catch and value. For the remaining 430-odd, there is not enough money to regularly assess the stocks. Some form of community-based management would certainly be desirable.

FURTHER READING

The classic work on community-based fisheries management. Johannes R E. 1981. The words of the lagoon: Fishing and marine lore in the Palau district of Micronesia. Berkeley: University of California Press.

Huxley's often criticized address. Huxley T H. 1884. Inaugural address. Fisheries Exhibition Literature 4: 1–22.

The Lofoten fishery in Norway. Hannesson R, Salvanes K G, and Squires D. 2010. Technological change and the tragedy of the commons: The Lofoten fishery over hundred and thirty years. Land Economics 86.4: 746–65.

An overview of fisheries management and Japanese cooperatives. Atkinson C E. 1988. Fisheries management: An historical overview. Marine Fisheries Review 50: 1–23.

Description of Japanese cooperatives. Makino M and Matsuda H. 2005. Co-management in Japanese coastal fisheries: Institutional features and transaction costs. Mar Policy 29: 441–50.

More on the Japanese cooperative system. Delaney A E. 2015. Japanese fishing cooperative associations: Governance in an era of consolidation. In: S Jentoft and R Chuenpagdee (eds), Interactive governance for small-scale fisheries. Cham, Switzerland: Springer, 263–80.

Description of Mekong River fisheries. Baird IG. 2007. Local ecological knowledge and small-scale freshwater fisheries management in the Mekong River in Southern Laos. In: N Haggan N, B Neis B, and I G Baird I G (eds)Fishers' Knowledge in Fisheries Science and Management. UNESCO, Paris: 247–66.

More on the Mekong. Hortle KG. 2007. Consumption and the yield of fish and other aquatic animals from the Lower Mekong Basin. MRC Technical Paper 16: 1–88.

Introduction to the Chilean Caleta system. Castilla J C and Fernández M. 1998. Small-scale benthic fishes in Chile: On co-management and sustainable use of benthic invertebrates. Ecol Appl 8 (Supplement): S124–S132.

A description of the Juan Fernández community management. Ernst B, Manríquez P, Orensanz J, et al. 2010. Strengthening of a traditional territorial tenure system through protagonism in monitoring activities by lobster fishermen from the Juan Fernández Islands, Chile. Bull Mar Sci 86: 315–38.

CHAPTER 4

Who Gets to Fish?

I n the beginning, anyone who wanted to fish could go fishing—then came "the tragedy of the commons."

The Commons

"The Tragedy of the Commons," a 1968 Science paper by ecologist Garrett Hardin,[1] has become the most important concept in managing fisheries. Somewhat simplified, it states that when a resource such as a fishery is held in common, everyone will act in their own best interest and, as long as it is profitable, will fish hard enough to deplete the resource. More will want to fish, and those already fishing will buy bigger or more boats, because if they do not catch the fish, someone else will until the stocks are depleted and profits disappear. Governments had already begun to see how the tragedy of the commons played out on the fishing grounds and had started to restrict fisheries at the same time the paper was published. Today, we broadly accept that fishing efforts need to be regulated.

Hot on the heels of regulation follows controversy. The most sensitive consequence of modern fisheries management is the effective privatization of the resource. As Hardin put it, "The tragedy of the commons as a food basket is averted by private property, or something formally like it."[2] In other words, exclusive access to a share of the resource prevents overexploitation in favor of maximizing its economic value. And that is the philosophical underpinning of Individual Transferable Quotas (ITQs), where individuals or companies have the government-issued privilege, usually revocable, to attempt to catch a portion of the allowable catch. Iceland and New Zealand have adopted this approach almost universally; other countries implement it partially.

Are there then no other options to manage a common resource than government control or privatization?

Elinor Ostrom, a political scientist, whom we have already met in the previous chapter spent much of her career showing that under certain conditions communities could, and had, managed commons for the collective good.

Ocean Recovery: a sustainable future for global fisheries? Ray Hilborn and Ulrike Hilborn, Oxford University Press (2019). © Ray Hilborn and Ulrike Hilborn 2019. DOI: 10.1093/oso/9780198839767.001.0001

The tragedy of the commons arises when it is difficult and costly to exclude potential users from common-pool resources that yield rational, utility-maximizing individuals rather than conserved for the benefit of all. Pessimism about the possibility of users voluntarily cooperating to prevent overuse has led to widespread central control of common-pool resources. But such control has itself frequently resulted in resource overuse. In practice, especially where they can communicate, users often develop rules that limit resource use and conserve resources.[3]

Using both real communities and laboratory experimentation, Ostrom showed that users will manage their common resources, if (1) most participants have tenure and plan to use the resource for a long time, (2) most participants share similar technologies, skills and cultural values, (3) the cost of communicating among participants is low, and (4) the cost of reaching binding and enforceable agreements is low. As long as the benefits outweigh the costs, the community will manage their resources effectively.

While possible, meeting these conditions is a rare event that is difficult to maintain over time. Let us examine what happened around the world once it became obvious that fisheries needed to be managed.

Once upon a time, "freedom of the seas" meant that anyone in possession of a fishing boat could fish anywhere in international waters. "Open access" in territorial waters of individual countries meant that anyone wanting to fish could do so, possibly having to pay a nominal fee for a permit. Consequently, 50 years ago, an unlimited number of people could enter any fishery, which led to a true tragedy of the commons. In 1968, Stan Rogers, a Canadian folk singer, wrote "Make and Break Harbor" about fishing for cod in Eastern Canada: *Foreign trawlers go by now with long seeing eyes, taking all where we seldom take any.*

The next 50 years brought various necessary government regulations that incrementally caused the creeping privatization of access to the once common resource.

The 200-Mile Economic Zone Is Declared: And the Race to Fish Is On

The most radical change was the declaration of 200-mile economic zones in the late 1970s. Most countries signed on and eventually the United Nations Law of the Sea ratified the 200-mile limits. Canada, for instance, said that everything within 200 miles would be theirs and foreign vessels could no longer fish there without Canada's permission. One could be forgiven for finding some irony in the fact that, notwithstanding the foreign vessel exclusion, Eastern Canadian cod did not recover, but collapsed catastrophically. However, the number of those who could fish in the 200-mile zone was now reduced to Canadians only.

There were high expectations that control over the 200-mile zone would make regulation of their fisheries more effective. As might have been foreseen, the results were mixed.

In the euphoria following the departure of foreign boats, but with "open access" still the rule, shipbuilding exploded, and by the mid-1980s, there were far more

Canadian fishing boats in Eastern Canada than there had been in the 1960s (foreign and Canadian combined). It did not take long for unlimited fleets to become untenable and "limited entry" replaced open access. Permits were no longer granted to anyone who wanted to fish, but limited permits were given only to those who could show a history of fishing. The permits could be transferable or nontransferable: that is, if a person failed to use them for a certain length of time, they were forfeit and the government would either retire them or reallocate them to a waiting list. More commonly, permits were transferable. The owner could sell them, and, quite often, they became very valuable. Because they were marketable, they created a new store of value in the fishery that could be taxed and used as collateral. Such limited-entry permits in the Bristol Bay sockeye fishery traded for $250,000 in 1996.

Limited Entry

Limited entry was another step to control the number of people allowed to fish to those who either won the lottery from a waiting list or could find money to buy a permit.

In theory, this should have done the trick. Human ingenuity and greed, however, have a way of subverting the best intentions. Whenever the number of boats was limited, fishermen put in more hours, added bigger motors, bigger boats, more crew, used better or larger nets and more electronics, and once again created too much fishing power. Governments noticed and countered with restricting the size of boats and the amount of net. When that did not bring fishing pressure down enough, boats were given a limit on either the number of days at sea per year they could fish or the amount of fish that could be landed on each trip. In the end, total-catch quotas became the management tools for many countries. Once a target catch was reached, the fishery would simply shut down.

The race to fish was on. In an "Olympic" system of a finite catch quota, each boat rushed to catch as many fish as possible before the fishery closed. Soon enough, boats that are more efficient were designed, and they hit the water.

It became apparent that too many boats were chasing too few fish, but if governments were to reduce the quotas sufficiently to start rebuilding the depleted stocks, it would cause the fishermen deep financial distress. In some cases governments decided to buy back permits and boats, but, "penny wise and pound foolish," they almost always bought the least efficient boats and left the catching power of the fleet almost unchanged.

The Next Holy Grail: Individual Transferable Quotas (ITQs)

Something new had to be done. Economists, armed with the logic of Garrett Hardin, weighed in, and the ITQ was born. With dedicated access to a fixed share of the catch, having a bigger boat was no longer necessary. The only way to make more money was to get a higher price for your fish or reduce your costs, and once retirement beckoned, you could sell your share to someone else. Both fleet size and total operating costs would come down.

Thus, the commons and its incentives to overfish faded into the past. In an ideal ITQ system, each vessel owner acquires an asset, the restricted access to a share of the catch, and each will be keen to grow the value of that asset. The fishermen and women would be stewards, rather than users, of the resource.

A very big benefit of ITQs is safety at sea. The race to fish and increasing fishing effort under the "Olympic" system had forced managers to cut down fishing seasons sometimes to just days or even hours. Because the start of the season had to be announced ahead of time, fishermen often faced the choice of risking their lives in a storm or missing the season altogether. In the Alaska crab fishery, the subject of the TV series "Deadliest Catch," an average of seven people died each year. In an ITQ system, you can choose when and where to catch your quota. As a result, there was only one death in the crab fishery during the next 6 years.

Market forces were meant to drive the reduction in fleet size without government-funded buy back schemes. Inefficient operators would sell their share to efficient operators. Ideally, some management authority could be devolved to quota owners because they would have strong incentives to collect reliable data on stock status. Some imagined that the management of the fishery itself could potentially be turned over to the quota owners with minimal government supervision. And with that, the ultimate solution for the problem of the commons, total privatization, would be reached.

The ITQ system can also be used to transfer fish catch between user groups by market mechanisms or government funding rather than by pre-emptive expropriation. For example, in 1974, US courts ruled that Native Americans in Puget Sound had treaty rights to half the region's salmon catch. This legal ruling stripped the nonnative commercial and recreational industries of half the fish they had been catching without any compensation. In contrast, when New Zealand courts ruled that the Maori had rights to much of the commercial harvest, the government bought large blocks of quota on the market (mostly through the purchase of a single large fishing company) and handed ownership to the Maori. The commercial industry felt no pain since there was no change to the quota.

A perennial and viciously contentious fisheries issue for many developed countries is the allocation of fish between recreational and commercial fishing. This is clearly a political, not a scientific decision, and, as a general rule, recreational fishing groups will win political battles because there are more of them by far than there are people involved in commercial fishing. A ballot initiative supported by recreational fishing groups in Florida banned the use of commercial fishing nets and thereby effectively eliminated most commercial fishing in the state. Through ITQs, if legislation allows, fish can be reallocated to recreational fishing through the market rather than the ballot or legislature. In the Alaska halibut fishery, recreational charter boat operators can purchase quota from commercial fishermen so their clients can catch more fish.

So Much for the Theory, but What Does the Reality of Fishing Under ITQs Look Like?

ITQs have been a boon for some and bust for others. As a one-size-fits-all remedy, they have not worked quite the way management agencies envisioned it. It is the human factor and the details and design of each system that will shape and determine the outcome.

Halibut: Dean Adams

Dean Adams had been a student in one of my classes at University of Washington, and he completed a Master's degree in our department, but his fishing career had started long before when he was 16 and a greenhorn crewmember on his uncle's halibut longliner in Alaska. His autobiography, *Four-Thousand Hooks*,[4] describes his first summer and ends when Dean and the crew abandon his uncle's boat for a life raft as the boat sinks with the holds full and the decks covered with halibut. He went on to become a very successful fisherman, longlining for halibut and black cod, who recognized the problems of the short halibut season that forced him to freeze all his catch because the market had been flooded with all the fish on the market at once. The open-access halibut fishery had moved from a 110-day season with 211 boats in 1970 to a 2-day season with 2,400 boats in 1989 and lives had been lost when the opening coincided with a major storm. Something had to change.

As a leader of the Seattle-based group of halibut fishermen, Dean lobbied successfully for an ITQ system for halibut and black cod similar to one that was already in place in Canada. In 1993, the fishermen were granted a share of the catch proportional to their share in recent years. Because of his excellent history of catch, Dean was allocated more black cod quota than anyone else was. The years after the ITQ system came in were great, catch quotas were high; fishermen could choose when to catch their quota; and the season was extended to 6 months. No longer did halibut need to be frozen but were sold over 6 months to the much more valuable fresh market for more money per pound landed. More importantly, fishermen could stay in port in foul weather, knowing they could catch their quota later. The fleet also changed. When half of the 2,400 boats sold their mostly small quotas, the fishery became much more professional with full-time jobs for the crew.

Dean continued to fish for a number of years; the value of his halibut and black cod quota kept rising while the hard work of fishing remained just as hard as ever and he was well past 16. With a comfortable retirement assured, he sold the boat and quota and began to write his book.

Mixed-Fisheries Dragging: Curley Brown

Ian (Curley) Brown's home port is New Plymouth, a typical small fishing town that supports a handful of boats and a small fish processing plant on the West coast of the North Island of New Zealand, a 4 to 5 hour drive from the two big cities,

Auckland to the north and Wellington to the south. Curley is a dragger; he fishes with trawl nets that are dragged along the bottom, catching a mix of species. Since New Zealand has an ITQ system for almost all fish species, anyone in a mixed-species fishery needs quota for each species caught. Curley owns quota for only about one-third of what he catches and leases the rest. The fish processors own most of the New Zealand quota, and the increasing lease fees will eventually drive Curley out of business.

Quotas are valuable commodities, and purchase prices are typically 5 to 10 times the value of annual landings. Lease fees may be over half of the value at the dock. Unsurprisingly, they are typically bought up by those who have the most access to capital, either fish processing companies or wealthy investors. The lease fees on quota are determined by market pressures, and because fishermen compete with each other to lease quota from owners, the fees rise to the price that efficient fishermen will pay. The less efficient will lose money leasing quotas.

Curley is the most efficient dragger in New Plymouth, but he will also be the last. One of his most vexing challenges is quota for snapper. Snapper are abundant close to shore, but the big fishing companies own most of the quota. Snapper is a highly prized fish; the quota is expensive; and, if he catches one snapper over his quota, he must stop fishing or pay a big penalty. Snapper is his "choke species." To avoid catching too many, he fishes farther out to sea where he knows how to find terakihi and gurnard even though they are less abundant there, which drives up his cost. He says that there will never be another dragger to replace him in New Plymouth because his 30 years of fishing knowledge allow him to barely stay in business for now, but no one else could make a living the way he does because they do not have his knowledge. His retirement will spell the end of trawling in remote New Plymouth. Those who fish closer to major markets will be the ones leasing the quota.

Longlining: Dave Moore

Dave Moore, on the other hand, has done very well under the New Zealand ITQ system. Dave is a longliner, based in Leigh, about an hour north of Auckland. He owns six longliners that fish around the top end of New Zealand. He leases his entire quota from processors. Part of his success is due to his partnership with a local small processor, Lee Fish, run by Greg Bishop, that is the crown jewel of New Zealand processors. They buy only high-quality, longline-caught fish and ship it whole, ungutted, in gel packs to markets around the world. Each box of fish leaving Lee Fish is labeled with the name of the boat, the skipper, and the date the fish were caught and can be on the plane to overseas destinations on the next afternoon. Because of their high value, Dave Moore gets a good price and can just afford the lease fees for the quota. He does not see how quota could stay with fishermen since quotas will always gravitate to corporate money. Dave has problems leasing quota, which he believes is too expensive.

Fishing in New Zealand would seem a great career path since no formal qualifications are required, and after a few years working as a crewmember, a skilled and ambitious person can become skipper of one of Dave's boats and make something like $150,000 per year. But fishing, especially longline fishing, is hard and dangerous work, 10 hours or more a day on a wet, cold, and heaving deck. Not many New Zealanders are lining up for it.

Have ITQ Systems Lived Up to Their Promise of Economic Efficiency, Safety, and Stewardship?

Here I must rely on personal knowledge and reading since there are almost no comprehensive reviews of how ITQ systems have worked out. There is no question that fisheries have become more economically efficient. Fleet size usually shrinks to the most efficient vessels, and, in many cases, the value of the product has risen greatly by, for instance, switching to the fresh market for halibut and taking the time to find the best-sized fish. There are some individual studies on improvements in safety, but there are no global surveys yet.

The evidence of increased stewardship is not so clear. Chris Costello at the University of California Santa Barbara, a frequent collaborator, suggests that fish stock collapses were less common in ITQ fisheries than in other types of management.[5] Tim Essington, a colleague at the University of Washington, failed to find any difference in the biological status of fisheries managed with ITQs.[6] In my experience, there is a definite stewardship benefit in the early years of ITQ systems when the quota is owned by the fishermen. But, in systems that allow absentee ownership or leasing, once quota ownership drifts away from the people on the water, the stewardship connection weakens. In the extreme, when most of the quota is leased, the incentive for stewardship disappears, and if the economics falter, fishermen are likely to revert to the bad behavior of the open-access days, doing whatever it takes to stay in business. However, there are stewardship benefits to fleet consolidation because it is easier to monitor a small number of large boats than a large number of small boats. On-board observers, satellite tracking, and other enforcement systems are much easier to install for consolidated fisheries.

Rock Lobster: Daryl Sykes

Daryl Sykes, a former lobster fisherman, has managed the New Zealand Rock Lobster Industry Council for 30 years, working with fishermen and quota owners to make the fishery sustainable and profitable. "He is the best fishery manager I have ever met," is how he is generally described. Initially, the industry organization had taken a very active role in management under his leadership, collecting the data to determine the quota and initiated the use of a management system that sets quotas based on changes in the catch per pot. When fishermen believed the quotas were too high, Daryl arranged a voluntary agreement to "shelve" some of the quota, that is to have all quota owners agree to use only a portion of it by retiring an amount of catching rights for one or more seasons. To identify the issues in data collection and management and solve problems, as well as manage people, has been his job for 3 decades.

But changes in ITQ ownership are making that job more challenging. In the early days of the ITQ program, the quota was owned by fishermen who could see the issues on the water, had newfound wealth in the value of their ITQ, and actively sought opportunities to enhance the value of their asset. Increasingly, lobster ITQs that now sell for about NZ $1 million per ton are owned by vertically integrated fishing companies who borrowed money to buy quota and need to pay their loans and earn a dividend on their investments. Indebted investors can be much harder to convince to shelve quota than the original fishermen, but over the past decade, they have stepped up support for industry initiatives, realizing that their equity is tightly bound to stock abundance.

Sector Allocation

There are other forms of allocation under the general classification of catch shares that are similar to ITQs. In the United States, for instance, a share of the catch is often assigned to some portion of the fishing fleet, often boats using the same gear or from the same port. Members of that "sector" can then deal with internal allocation as they wish. Sector allocation has been very successful in the Alaska pollock fishery, which, by volume, is the largest fishery in the United States. There are three sectors: the boats that land fish at shore-based plants, the factory catcher boats, and motherships—floating processors that do not catch their own fish. While these are not legally ITQ fisheries, they operate in practice very much like an ITQ, except the allocation between vessels is done by the members rather than the government. Let me introduce some of the dominant figures in that fishery.

The Legend: Chuck Bundrant

Chuck Bundrant, a fishing legend, is the founder and largest shareholder of Trident Seafoods, the biggest fishing company in the United States. As a teenager, he drove from his native Tennessee to Seattle and looked for work on a fishing boat. Starting as a deckhand, he became one of the most competitive and successful fishermen and processors. I met him in Bristol Bay in the late 1980s. We talked about my work in New Zealand, and I asked him what he thought about an ITQ system for some of his fisheries and got an unequivocal response. "That isn't fishing, if it happens in Alaska I am getting out of this business." As we'll see, he changed his mind.

Alaska Pollock: Bernt Bodal

Bernt Bodal came to the United States from Norway, also as a deckhand who worked his way up and became majority owner and CEO of American Seafoods, a company that owns six factory ships in Alaska. In the mid-1980s, the Alaska pollock fishery was run as an Olympic system. Several-dozen factory ships would race each other to catch the most fish before the catch quota was used up and their season was over. I have been told that in the mid-1990s, Bernt commissioned an economist to see how the company's bottom line would improve if, instead of racing to fish in an Olympic system, each company had a fixed share of the catch and could catch fish over a longer season. The economist crunched the numbers and came back with one number for his estimate of increased profits on a single 3×5 card. This was a very large number and was ultimately realized when the 1998 *American Fisheries Act* and the cooperative agreement among the factory ships to share the catch was born.

More Alaska Pollock: Al Chaffee

Another self-made fishing company owner who profited from this kind of allocation is Al Chaffee. He started his career working for a group of fishermen, and then acquired a small pollock factory trawler, the *Highland Light*. We usually see each other in the summer in Bristol Bay when he comes back to operate his old salmon processing plant for the new owners. In the early 2000s, when the Bristol Bay salmon fisheries were suffering, I would ask Al how his pollock boat was doing, and each year he would sigh and say that he could not believe how much money could be made in pollock. In 2008, the point came when "they offer you so much money you just can't refuse," and he sold his boat and the catch shares to American Seafoods. The *Highland Light* retired to a Seattle dock, and American Seafood Group's other ships caught Al Chafee's former share.

The pollock factory ship cooperative agreement rapidly spread to the other sectors of pollock and has become the model for many other forms of sector allocation elsewhere in the United States, especially New England. As opposed to ITQs, the government determines the total-catch share for each sector rather than for any individual. Those who fish in each sector decide the internal allocations. In theory, a sector could maintain an internal Olympic system and race each other to fish. In practice, sectors usually allocate based on historical catches and enforce them with legally binding agreements that allow trading allocations within the sector.

So Who Does Get to Fish under ITQ Systems?

Initially those who were fishing and had a sufficient history of catch were allocated ITQs. But as time passes, unless the system is designed to avoid it, consolidation of quota ownership seems inevitable. The Bernt Bodals buy out the Al Chaffees, and fewer and fewer individuals and companies own the quota. Large-ship fisheries, companies like Trident and American Seafoods, control an increasing portion of the quota and fish it with their company boats. For smaller-boat fisheries like those in New Zealand, where Curly Brown and Dave Moore fish, those who get to fish are those who can compete in the market to lease quota.

Obviously, there is a problem here.

Economists Dream of ITQs and Catch Shares, but Other Social Scientists Have Nightmares

Let us look at social equity. In essence, ITQ systems give away access to public property. The vast wealth that was solely generated by this limited access to such fisheries as New Zealand Rock lobster and Alaska pollock could not have been generated through open-access fisheries. A few economists, among them the Canadian Parzival Copes, were prescient enough not to share the dream and foresaw the outcome. Copes warned that valuable quotas would be consolidated, fishing itself would become concentrated geographically and financially, jobs would be lost, the path from deckhand to fleet owner would no longer be possible, and communities who depend on their fish stocks would see their resource slipping into outsiders' control, and their very existence would be under threat. And so it came to pass. The economic efficiencies of ITQ systems had social consequences that many do not like.

The Mighty Geoduck

Copes and his like-minded few do have alternatives. They proposed that rather than assigning quota privileges, governments should sell them at auction, thereby capturing the value of the fishery for the public. The only American example I know of is the annual Washington State auction for geoduck quota. Geoduck, pronounced *gooey duck*, is neither gooey nor a duck but a bivalve with an excessively long, rather unprepossessing siphon. Prized in Asia, it is a high-value fishery. However, as Copes might have foreseen, only the processors have the money to bid on the quota. The divers, who flush out the geoduck with pressure hoses, are contract employees but can rarely aspire to purchasing harvest rights much like Curley Brown and Dave Moore do in New Zealand when they lease quota.

This consolidation of ownership and threats to small communities is a legitimate concern. Evelyn Pinkerton and Courtney Carothers, two social scientists who study the impacts of ITQs and other allocation methods on small communities, small-scale fishermen, and fishing crew, have documented the decline in quota ownership in small communities and the aggregation of fishing rights. Whether it is limited-entry permits, ITQs, sector allocation, or auction, putting a price on the right to fish means that whoever has access to money wins, whereas small operators, especially aboriginals and those from very small communities, are unlikely to be able to compete for fishing permits or quota.

Nonetheless, there is increasing pushback against ITQs and catch shares, and I think it unlikely we will see expansion of those programs in the United States or in Europe. In Iceland, small-boat fishermen have been attacking the "cod barons" who own most of the cod ITQ, and some political parties have indicated a willingness to either take back some of that quota or start taxing it heavily.

Tasmanian Abalone

In a more equitable system, how can we retain the economic and safety benefits of catch shares? Aggregation limits, where no one can own more than a certain percentage of quota, have been tried and have proved difficult to enforce. When the price for ITQs rose into the stratosphere in the Tasmanian abalone fishery, for instance, processors were willing to lend the money to potential fishermen, provided that they delivered all abalone to their plant. Given the weighty debt, the fishermen found themselves owned by the processing company. Other catch-share systems have required quota owners to have been "on the water," to have fished in the quota's fishery, before becoming a nonfishing investor.

And There Are Still Other Ways of Allocating Fish

In the European Union: Producer Organizations

Within the European Union, individual countries are granted a share of the total quota for each stock, and each country then allocates its quota to individual "producer organizations," which are voluntary associations of fishermen, in theory, much like sectors in the United States. The producer organizations internally distribute the catch and determine membership. So if the producer organization lets people sell their catch internally, we would expect a *de facto* ITQ system. But if the producer organization had a "use it or lose it" policy, ITQ-type concentrations would not happen.

What About Community-Based Allocation for Small-Scale Fisheries?

The community-based approach grants fishing rights to communities or organizations of fishermen who then decide both who can fish and allocate the rights. In theory, the sector allocations of Alaska and New England also are community based, but since each member is given an individual tradeable share, the sector invariably devolves into a *de facto* ITQ system.

In Chile, there are hundreds of fishing communities, called Caletas. Each organization has its own way of allocating the fishing rights and benefits among members. Each union has its own regulations. Some will allocate rights among divers only; others among all members (fishers and divers); and others will have restrictions (temporary cancelling rights) for members who violate internal regulations. Any community-based approach will be governed by the political dynamics of the membership. Powerful families or groups of families can dominate a community and naturally favor their members.

Finally, Foreign Fleets

Originally, the many islands scattered across the Pacific had their own systems of community-based fisheries allocations. With the advent of the 200-mile exclusion zone, they saw their chance of steady income by leasing the rights to fish to foreign vessels in exchange for royalties. Many Western Pacific countries' national budgets come from payments for access to the valuable tuna resources. The UK administration of the Falkland Islands has allowed foreign fleets to catch squid around the islands in exchange for significant lease fees. Many African countries have also licensed foreign fleets to fish in their waters in exchange for either royalties or various forms of aid.

Almost all coastal nations have adopted some legal framework to exclude foreigners from fishing in their waters. The ways and means of regulating their own citizens are in a permanent state of flux as to who gets to fish changes continually. Countries need to make decisions about whether fisheries should be managed for economic efficiency, maintenance of access to fishing for small and isolated communities, or for other social objectives. Countries or regions also have to decide how much to allocate to commercial or recreational uses. New Zealand's ITQ program was specifically designed for economic efficiency; Alaska's system of salmon fishing permits is designed to maintain single owner-operators. There is no "best" approach to allocating fishing, and there is no scientific answer. It all depends on what the government and society want to achieve with their fishery. But there is no question that the Western tradition of open access is rarely an approach that will produce the best biological, social, or economic outcomes.

On the High Seas, Only Rich Nations Get To Play

High seas fisheries are governed by Regional Fisheries Management Organizations (RFMO). There are over a dozen. Each one has jurisdiction over specific species of fish in a defined area. Five deal with tuna fisheries; one with Antarctic fisheries; one with whaling; and many others with specific fisheries in different parts of the ocean. Each one has its own approach to regulating access.

Beyond the 200-mile zones, open access had been the traditional rule; countries or companies that could build boats to fish the high seas were the ones that caught the fish. Rich countries such as Japan, Spain, Korea, the United States, and China dominated early high seas fishing. Today RFMOs allocate catch based largely on historical catches with the result that the same countries continue to dominate the high seas fisheries.

• •

FURTHER READING

An early analysis of the consequences of open-access fisheries. Gordon H S 1954. Economic theory of a common property resources: The fishery. Journal of Political Economy 62: 124–42.

The classic paper on tragedy of the commons. Hardin G. 1968. The tragedy of the commons: The population problem has no technical solution; it requires a fundamental extension in morality. Science 162: 1243–8.

One of Ostrom's works on how communities can manage resources without privatization. Ostrom E. 2008. Tragedy of the commons. In: S N Durlauf and E L Blume (eds), The New Palgrave Dictionary of Economics, 2nd edition. Basingstoke: Palgrave MacMillan.

Dean Adams's story of his first year as a fisherman. Adams D J. 2012. Four thousand hooks: A true story of fishing and coming of age on the high seas of Alaska. Seattle: University of Washington Press.

An economist's critique of ITQ systems. Copes P. 1986. A critical review of the individual quota as a device in fisheries management. Land Economics 62: 278–91.

A critique of most forms of limiting entry to fisheries and its impact on small communities. Lowe M E and Carothers C. 2008. Enclosing the fisheries: People, places and power. Bethesda, MD: American Fisheries Society.

A critique of the impact of quota leasing under ITQ systems. Pinkerton E, and Edwards D N. 2009. The elephant in the room: The hidden costs of leasing individual transferable fishing quotas. Marine Policy 33: 707–13.

CHAPTER 5

The Global Status of Fisheries

a long tale of scientists, opinions, and papers written and refuted, all in the pursuit of the same truth

Gloom and Doom

All Fisheries Will Be Collapsed By 2048!

So said the front pages of *The New York Times* and the *Washington Post*, even the *BBC* evening news covered it on November 3, 2006. Front-page headlines about fisheries are rarer than hen's teeth, unless of course it is really bad news. And so it seemed to reporters covering a *Science* magazine paper by Boris Worm and thirteen others, stating that, if current trends continue, all commercially exploited fish stocks would collapse by 2048.[1]

Was it true? Not exactly, and certainly not regarding a total collapse by all. In the fisheries world, the decline of fisheries was hardly hot-off-the-press news, but it had long been a concern. There had been reams of scientific papers and coverage by popular media about it for at least a decade. Don Ludwig, Carl Walters, and I had published an article in 1993,[2] also in *Science*, warning that fisheries managers need to act before stocks are in trouble, and that waiting for scientific certainty about fish stocks was a prescription for increased overfishing. I am quite proud of this paper: it is my most cited one, and it too was covered by *The New York Times*, even with a picture of the three of us. Needless to say, we did not make the front page, but were relegated to the back.

When Canada closed the Newfoundland cod fishery in 1992, the story of global fish stock collapse received its greatest boost. Cod had been the backbone of Newfoundland's economy for 500 years. In the 1960s, foreign fleets had plundered the stock, but they were evicted by the 200-mile zone declaration in 1977. Canada controlled the fishery now, and its managers were supposed to rebuild it. It obviously did not work out that way. The closure in 1992 was a terrible blow to the fishermen, but it was hoped that it would take only a year or two for the stock to recover and they could go fishing again; 25 years later, there are signs of rebuilding but not enough yet to allow any significant fishery.

Ocean Recovery: a sustainable future for global fisheries? Ray Hilborn and Ulrike Hilborn, Oxford University Press (2019). © Ray Hilborn and Ulrike Hilborn 2019. DOI: 10.1093/oso/9780198839767.001.0001

Cod is the poster child of fisheries mismanagement, and Mark Kurlansky is its biographer. In *Cod: A Biography of the Fish That Changed the World*,[3] he tells us how important cod was to the social fabric and economic strength of Newfoundland, Iceland, and the State of Massachusetts. In the 1600s, one could "walk across the backs" of cod, they were so plentiful. Then came the long, slow decline in the 1990s, not only in Newfoundland but all around the North Atlantic.

There had been trouble elsewhere too: in my lifetime, the California sardine collapsed in the 1950s, the Peruvian anchoveta in 1970, and the North Sea herring in the late 1970s. While none of this was a secret in the fisheries community, only the iconic Newfoundland cod attracted global media attention.

Speaking of icons, let us talk about tuna, specifically about the magnificent, silvery, torpedo-shaped, and oh so tasty and widely hunted bluefin tuna. If ever there was a media star, it is the lightning fast, predatory bluefin.

In 2003, Ransom Myers and Boris Worm estimated in *Rapid Worldwide Depletion of Predator Fish Communities* that "the large predatory fish biomass today is only about 10% of pre-industrial levels." This paper got a lot of publicity, and it implanted in the public mind that most of the large fish of the oceans are gone, a topic that has been hotly discussed ever since. Atlantic bluefin tuna is not only on the International Union for the Conservation of Nature red list as endangered, but is also on every consumer guide's equally red do-not-eat list. Pacific bluefin tuna is listed as vulnerable, and the southern bluefin tuna of the Indian Ocean as critically endangered. All of them are in the spotlight of conservation organizations and the public interest. The star and example of all that is wrong with fisheries in *The End of the Line*, Charles Glover's 2004 book and subsequent documentary, is none other than our icon, the Atlantic bluefin tuna.

So it was no real surprise that the claim of a demise of all fish by 2048 would reinforce the general public's opinion that fishing was draining the ocean of fish.

I have to admit that the boldness of the claim shocked many of us who work in fisheries. After all, I had spent almost all my career working on fish stocks in the Pacific Ocean. I started in Western Canada, and then moved on to work on Pacific tuna, then to fisheries in Australia, the Western United States, Alaska, and New Zealand. Few of the fisheries I knew well were destined for collapse. On the contrary, most were managed so well they would be rather healthier in 2048. Uneasy in my mind, I dug into the data analysis and got worried.

What was the definition of *collapse* in this paper? For Worm's group, a collapse happened if the catch in any year was less than 10 percent of the highest catch observed. The data came from reported landings by country and type of fish, compiled by FAO for roughly 20,000 country/fish type combinations. Plotting the collapsed stocks against time, they found a downward trend with about 30 percent of stocks collapsed in 2006. Projecting forward, the curve would reach 100 percent in 2048.

What do we mean by *fish stock*? A *stock* describes the fish in any particular region that belong to the same species. In theory, each stock should be a breeding unit that is geographically separated from the same species in other areas. In real life, the boundaries of stocks are often drawn quite arbitrarily based on national boundaries or geographic features. The stocks in the Worm 2006 paper were drawn by country and species or species group.

My uneasiness rose when I looked in detail at stocks listed as collapsed on the West Coast of the United States and in New Zealand, fisheries I know well. I made a list of the collapses and sent it to colleagues who were very familiar with the fish stocks in question. One of them was David Gilbert, from NIWA, a New Zealand government fisheries laboratory, and he wrote in an email to me:

> This leaves 69 stocks of which 10 have "collapsed" by Worm's criterion. But the collapsed stocks are almost all cases where landings have been partially separated into a collapsed and another group. Of the 10 "collapsed" stocks the only genuine candidate is the dredge oyster which has been decimated by a bonamia parasite epidemic. Worm's analysis is therefore completely fatuous.

The problem here is a matter of lumpers and splitters. In the 1950s, many species were lumped together with their related species, but later, when data collection improved, these lumped groups were split into individual species so the lumped group had almost no reported catch. But in Worm's analysis, these lumped groups were still considered collapsed.

On the US West Coast, Alec McCall of the US National Marine Fisheries Service identified the same problem and added another category, the *false collapse*. When Mexico declared its 200-mile limit, US boats were no longer allowed to fish there. The data, however, go back to pre-200-mile days. Consequently, the species US boats had fished for suddenly showed up as zero in catch reports from the United States—and became the false collapses in the Worm Group paper.

Several critiques by fisheries scientists, including mine, were subsequently published in *Science*, and a few weeks later, Boris Worm and I were invited to discuss the issue on NPR, the National Public Radio in the United States. Boris and I did not know each other, but I knew he was born and educated in Germany as a marine ecologist and then moved to Halifax in Canada. That meant he worked on the fisheries of the North Atlantic that were widely overfished at the time. It became obvious during the NPR discussion that Boris and I could talk to each other as scientists. Many emails followed to understand why we had such different perspectives.

Was it because of where we worked? Boris worked on the poorly managed North Atlantic fisheries, I work on the much more sustainable fisheries of the Pacific. Or was it because we differed on what fisheries are about? I, and most scientists who work in fisheries management, consider fisheries a way to produce food, and we accept that fishing will change ecosystems. Boris, a marine ecologist, would more likely think of fisheries as a threat to marine ecosystems. Where I see a necessary change in the ecosystem to produce food, marine ecologists might see degradation. Then again, might it be where our data on fish stocks come from? Marine ecologists tend to work close to shore, in places that can be reached with small boats. Fisheries scientists like me rely on large government research vessels because the major fisheries

A Closer Look

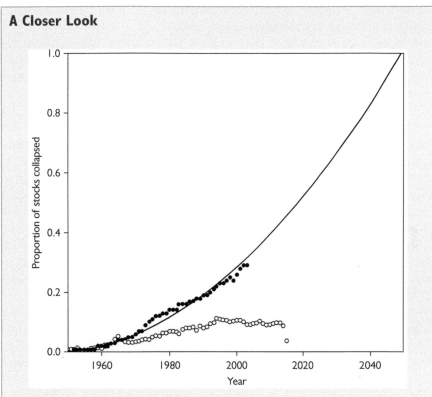

Figure 5.1 This figure shows the proportion of taxa classified as collapsed in the Worm 2006 paper from 1950 to 2003 (the solid dots), and the projection from that paper that suggested all stocks would be collapsed by 2048 (the solid line). The open circles are the fraction of stocks collapsed in the RAM Legacy Stock Assessment Database.

The projection to 2048 is obviously highly uncertain, but the data did show a clear trend in an increasing proportion of stocks being classified as collapsed. However, when we look at the abundance data for the assessed stocks of the world in the RAM Legacy Stock Assessment Database, we see the trend in the proportion of stocks collapsed (open circles)—a much lower fraction of stocks were collapsed—and the trend leveled off in the 1990s and has declined slightly since then. The RAM Legacy database does not have data for the major fisheries of South East Asia, and we expect stock collapse to be common in that region. The open circles show why those of us working in fisheries management were skeptical of the claim that all fish stocks would be collapsed by 2048. In the places where most of us worked (places with abundance estimates), the trend in collapse was not increasing, and the fraction of stocks collapsed was not as high as reported in the 2006 paper.

are out in the open ocean. In general, the coastal zones show much heavier impacts from nearshore fisheries and human-made effects.

How could we overcome our differences? A grant from the National Center for Ecological Analysis and Synthesis (NCEAS) in Santa Barbara, California, allowed us to assemble a group of roughly twenty scientists who would meet three times over 2 years to attempt to understand what was really going on in world fisheries. We agreed at the outset that we needed to look at data on the abundance of fish stocks, not reported catch data. Abundance data come from research surveys that estimate trends by using either fishing gear in a scientifically designed pattern or acoustic methods (sonar). Acoustics are usually combined with some fishing to identify the species, measure each fish's length and take out its ear bones to determine its age. Fishing gear is generally used for bottom-dwelling fish, and acoustics are used for fish that swim in midwater.

Scientific surveys should cover the entire range of the stock, including areas where fish are rare. Even though it irritates fishermen when research vessels spend time and money sampling where they do not find fish, a survey worth its effort and money must sample beyond the range of a stock to give us a true idea of abundance.

When all available data from research surveys and landed catches are taken together, the heavy lifting starts. It is time for fisheries scientists to work on *stock assessments*. Without them, any manager trying to set limits is flying blind. With them, she or he is given eyes to estimate the all-important trends in abundance. Our NCEAS group looked at stock assessments of 166 major fish stocks. Each one told us about estimates of historical abundance, the reported catch, and the *recruitment*, that is, the number of fish born that survive to be counted and enter the fishery.

Meet the North Sea Herring, a Fine Example of Overfishing Followed by Rebuilding That Deserves a Graph to Tell Its Story

Herring, like sardines, are small pelagic fish of great aggregations. Pelagic fish swim in the water column, neither on top nor on the bottom, and many of them are boom-and-bust fish that often cycle wildly.

During WWII, the North Sea herring had been left pretty much alone, but by 1946, fishing started again. In the mid-1960s, just under 1.2 million tons were caught, yet the stock did not look all that good. Warning lights went up, stock and catch tumbled, and, in the late 1970s, there was no fishing at all. The herring recovered, and eventually there were some very good fishing years followed by a downturn in the 1990s when the herring were below the target and catches reduced. Presently, under EU management, stock and catch are kept on an even keel. Managing cyclical stocks is a never-ending challenge to smooth out the boom and bust.

A Closer Look

This graph shows the spawning population size (tons of mature fish) for the North Sea herring stock from 1947 to 2015 as solid dots. The open circles show the catch. After WWII the stock was very large, and about 700,000 tons were caught annually. However, the stock declined from over 5 million tons to only about 500,000 tons. Catch declined accordingly, and the fishery was effectively closed for several years in the late 1970s.

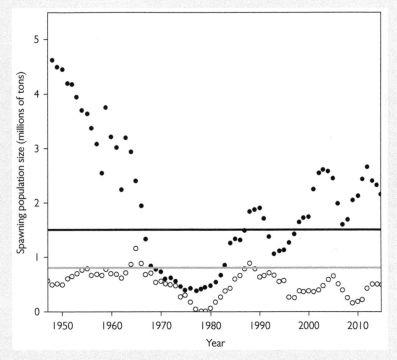

Figure 5.2 The horizontal solid line shows the current management target for the spawning population size (1.5 million tons), and the horizontal gray line is the management limit—if the spawning population size falls below this level (800,000) tons, the fishery will be closed. Since the late 1980s, the stock has been near or above the management target. Data from www.ramlegacy.org.

Maybe it Isn't all Gloom and Doom

Once we had assembled the data on trends in abundance from surveys and stock assessments, we published *Rebuilding Global Fisheries*[4] in *Science* in 2009. What we had found was that on average, our data showed no downward trend either in the survey data or in the stock assessment data.

More importantly, we found that in different regions, trends differed wildly. In Alaska and New Zealand, widespread overfishing never happened, and trends were stable. Elsewhere, including Iceland, Norway, and the US East Coast, many stocks

A Closer Look

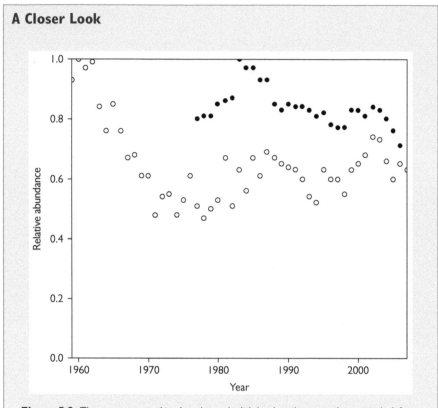

Figure 5.3 The average trend in abundance (solid dots) and surveys (open circles) from the 2009 paper by Worm and others. We found no overall downward trend.

were rebuilding after having been overfished. But there were still areas like the Gulf of Thailand that were overfished, and stocks declined. We concluded:

> In 5 of 10 well-studied ecosystems, the average exploitation rate has recently declined and for seven systems is now at or below the rate predicted to achieve maximum sustainable yield. Yet 63% of assessed fish stocks worldwide still require rebuilding, and even lower exploitation rates are needed to reverse the collapse of vulnerable species.

We found something to meet everyone's expectation; yes, there were areas where stocks were declining, and yes, there were areas where stocks had not been over-fished, or were increasing after overfishing.

Our NCEAS group was very excited about the results, and many in the fisheries management and conservation community were pleased to see the consensus reached between the supposed warring factions. Even *Science* weighed in with *Détente in the fisheries war,*[5] and wrote that *after a controversial projection that wild-caught fish would disappear, top researchers buried the hatchet to examine the status of fisheries—and what to do about it.* Boris was convinced of more page-one press coverage. I was skeptical—all fish gone by 2048 is front-page news; some fisheries doing well is not. There was a wager for a bottle of champagne. It was *The New York Times*, page 23, and I drank the champagne. More symbolically, Boris promised a seafood banquet on December 31, 2047, now confident that there would be fish to serve. I hope that my younger colleagues or I can hold him to that promise, because that day will be my hundredth birthday!

The two key results of our NCEAS work were the recognition by the marine ecology community (well, more than a few anyway) that many fisheries increased and were sustainably managed, and the development of a database of fisheries stock assessments.

We continue to update and expand this database and named it the RAM Legacy database in honor of Ransom (Ram) Myers, a colleague who developed the first large-scale database on fish assessments in the 1990s, but, sadly, died in 2007, just as we began our working group. In 2009, the database contained assessments of 166 stocks representing about 20 percent of global catch. Even though we called our paper *Rebuilding Global Fisheries*, we stretched the definition of *global* quite a bit because most of our data came from countries wealthy enough not only to conduct scientific research on fisheries but also make it public.

The RAM Legacy database is in the public domain (www.ramlegacy.org), managed at the University of Washington, and it has expanded to over 1,200 fish stocks that represent a little over 50 percent of global fish landings. It covers the fisheries of the United States, Canada, Norway, Iceland, the European Commission, Peru, Chile, Argentina, high seas tuna fisheries, New Zealand, and Japan, as well as most of the major fisheries of South Africa and Australia. Still missing are the key areas of large catches in South and South East Asia.

We can now speak with more authority about the status of fisheries in large parts of the world, always aware that we are still missing most Asian, African and Latin American fisheries with the exception of Peru, Chile, Argentina, South Africa, and North West Africa. While we do know quite a bit about those stocks, our knowledge does not come from scientific stock assessments.

What Have We Learned from All This Compilation of Data? Let Us See the Details

The RAM Legacy database[6] tells us that there are increases in abundance in Atlantic Ocean tunas, on the Canadian West Coast, in the European Union Atlantic fisheries, in Norway, Iceland, and New Zealand, and in all regions of the United States. In Canadian East Coast fisheries and Japan, increases in abundance proceed at rather glacial speeds. South America and the Mediterranean, alas, are suffering declines of low abundance without recovery in sight. Indian Ocean and Pacific tuna fisheries did decline but have not yet fallen below the target abundance. In general, stocks that were overfished in 2006 could increase if fishing pressure were reduced.

Now onto the numbers. Overall fish stock abundance in the areas where we have data increased by 12 percent between 2000 and 2012, and the proportion of stocks

A Closer Look

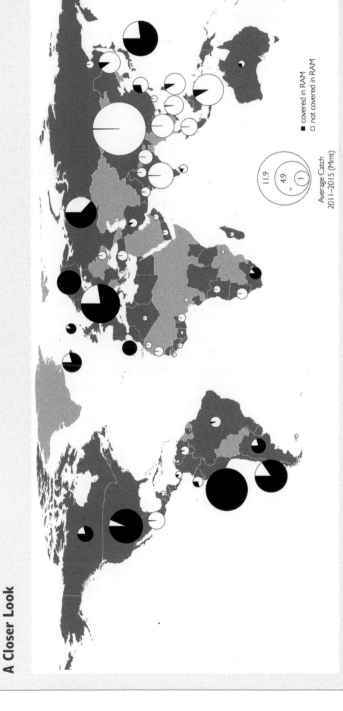

Figure 5.4 The amount of catch from different countries in the world (size of circle) and the amount of catch from assessed stocks in the RAM Legacy Stock Assessment Database (shaded region). South and Southeast Asia from India to Indonesia to China and Taiwan are the major fisheries of the world without scientific assessments.

- ■ covered in RAM
- □ not covered in RAM

11.9
4.9
1

Average Catch
2011–2015 (Mmt)

considered overfished decreased slightly from 23 percent to 21 percent. There are big regional differences in both status and trend. In 2000, the stocks of eight of eighteen regions were on average above biomass at maximum sustainable yield (BMSY), and in all other areas, stocks have, on average, increased except for South America, the Mediterranean, and West Africa. In the most recent years, the median stock status for the Mediterranean, South America, the Northwest Pacific, the US East Coast, South Africa, and the Canada East Coast is below BMSY.

Even if it should by now be pretty obvious, let us say it again, stop fishing so hard, and stocks will rebuild.

Whatever Happened to the Icons of Fisheries Failure, Cod, Tuna, Toothfish, and Orange Roughy?

Let us begin with what I call the Oh-my-God-what-have-we-done? chart of North Atlantic cod abundance that starts with the Portuguese fishing for cod far away from home and the times when the New Englanders fancied that you could walk across the cod's backs dry shod. What happened after that is what happens when everyone thinks that a resource is without limit.

Modern stock assessments for cod began in 1970 at an abundance of 8 million metric tons for all the major cod stocks (see Figure 5.5). The decline began in 1980 and became dire, particularly for Newfoundland, in the 1990s. Total overall abundance bottomed out in 2000 at 3 million metric tons, and it was time to reduce or stop fishing altogether. Contrary to much scare mongering, cod had not disappeared everywhere, but it decidedly needed a breather from harvesting. Recovery started and accelerated in 2005, driven by the very large Northeast Arctic stock found primarily in Northern Norway and European Russia. Icelandic cod too started to increase strongly. With new management under the EU, fishing has been generally good. The Western Atlantic stocks, on the other hand, did not do all that well. Recovery was painfully slow or did not happen at all. Even after fishing was closed, the Newfoundland stock did not budge for over a decade. Abundance started to inch up in 2005, and it is now estimated at 350,000 tons, still a very far cry from the millions of tons in the 1950s.

"What, You Eat Tuna? I Thought They Were All Extinct." Reproach from a Friend When My Wife Ordered Tuna Belly Sushi

Our media star, the bluefin tuna, became notorious in 2013, when Kiyoshi Kimura, the owner of a Japanese sushi restaurant chain, paid $1.76 million for one 489-pound Pacific bluefin. The year before, Kimura had paid $736,000 for a single fish.

A Closer Look

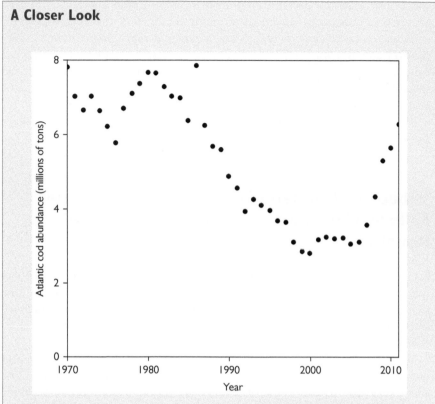

Figure 5.5 The total abundance of Atlantic cod from 1970 to 2012. Total abundance is now at about the same level it was in 1970, but many individual cod stocks remain in poor condition.

With that kind of publicity, it was no wonder that ocean conservationists rallied around the bluefin to prevent its likely extinction. Would it not pay fishermen to go to the end of the earth to catch the very last one?

It would not, because Mr. Kimura's almost annual crazy bid at the tuna auction is a publicity stunt to keep his restaurant patrons wondering if they are eating a piece of the most expensive fish in the world. And if he can convince them that they are, that will be the best piece of sushi they ever ate. The real price of bluefin tuna is much more mundane. While Mr. Kimura spent $3,600 per pound for his 2013 prize tuna, the next bidder probably got his bluefin at the same auction in Tokyo for the average price of $17 per pound.

In 2003, Myers and Worm concluded that 90 percent of the large predatory fish[7]—our tuna would be one of them—were gone by 1980. I have a soft spot for tuna because I worked on them from 1985 to 1987 at the South Pacific Commission (SPC) in New Caledonia, an international agency that does the science for the tuna resources of the countries of the Western Pacific, and my skeptical scientific hackles rose right away.

The Western Pacific is the mother lode of tuna and home turf of the small world of tuna scientists with whom I had become connected. I knew immediately that Myers and Worm could not be right. They had used the catch per hook of the Japanese longline fleet as an index of the abundance of different species, and in most cases, the catch per hook had declined rapidly and dramatically. But everyone who worked on tuna knew that there were hot spots for tuna where they were easily caught, which drove up the catch per hook initially, but the hot spots soon turned into regular fishing grounds with regular catches. However, over the entire range of tuna, the catch per hook had not declined nearly so much.

The tuna scientists were not amused by Myers and Worm. My favorite critique came from Carl Walters at the University of British Columbia with the title *Folly and Fantasy in the Analysis of Spatial Catch Rate Data*.[8] The journal *Nature* that had published Myers and Worm, took 2 years to publish an extended critique, and not until 2005, did it accept John Hampton's (SPC) article *Decline of Pacific Tuna Populations Exaggerated?*[9] As usual, disaster got the publicity, and the good news limping in way behind was barely noticed.

At the time of his paper's publication, I did call Ram Myers, an old friend and collaborator, and said, "Ram, this is impossible, if these stocks were depleted to 10% of the original abundance by 1980, how is it that global catch of these species increased three fold from 1980 to 2000." It seemed pretty persuasive that stocks could not have been depleted by small catches and still produced so much more catch in subsequent years.

The definitive status of tunas was published in 2011 in the Proceedings of the National Academy of Sciences by a group from Simon Fraser University in Canada led by Maria José Juan-Jorda.[10] Armed with the tuna assessments from international research groups like the SPC, they showed that globally tuna stocks had declined by 60 percent from 1954 to 2006. By 1980, the decline had only been 20 percent and not the 80 percent Myers and Worm claimed. An initial decline of 60 percent may strike a non-fishery reader as terrible, but it is precisely by how much we want a fished population to go down to get the best possible long-term sustainable yield. Most tuna stocks are now at or above the target levels.

But What about Our Media Star and Icon of Mismanagement, the Atlantic Bluefin?

There is no doubt that Atlantic bluefin was legally and illegally overfished. The key fishing countries, Japan, Spain, and France, eventually buckled to intense pressure from environmental NGOs, and by 2010, it seems that they stopped most illegal catches and considerably reduced overall harvest. In response, the abundance of bluefin in the large Eastern Atlantic population has increased nicely.

A Closer Look

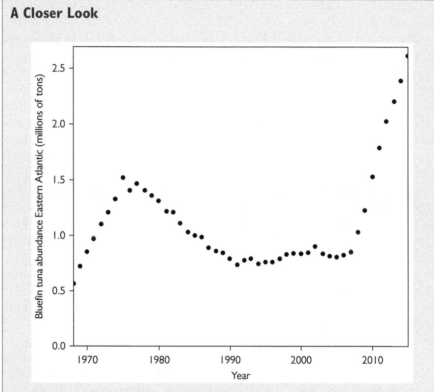

Figure 5.6 The estimated trend in abundance of Eastern Atlantic bluefin tuna. While abundance did decline in the 1980s, and there was lots of unreported catch, it was far from going extinct, and it has rebuilt rapidly since the late 2000s. It remains classified by IUCN as "endangered."

Patagonian Toothfish and Orange Roughy: Everyone Knows They Are Overfished

The other two notorious examples of unsustainable management are the Patagonian toothfish, also known as Chilean sea bass, from the Antarctic and sub-Antarctic Islands, and orange roughy from New Zealand.

Toothfish fishing began in earnest in the 1980s, and it was immediately embraced by the white-tablecloth establishments as an attractive flaky whitefish with high fat content. Just as quickly, it also appeared on the red list of most environmental NGOs because of extensive illegal fishing, bycatch of seabirds, and worries whether fish that live 60 to 70 years could even be managed in a sustainable way.

But the critics were proved wrong. The gold standard of fisheries certification is the London-based Marine Stewardship Council (MSC), an NGO founded initially by the World Wildlife Fund and Unilever, a major retailer. MSC certification is important in Northern Europe, where the major retailers will not sell fish without it.

Getting certified is not easy. Many boxes need to be ticked in an elaborate evaluation of the management system and the health of both the fish stock and the ecosystem.

A letter to me from MSC arrived in January 2003. Toothfish from South Georgia Island, a British sub-Antarctic territory, were determined by independent evaluators to meet their standard. Would I serve on an "objections panel?" Several environmental NGOs had filed objections. Via conference calls, our panel met with the proponent of certification, in this case the British Government, and the objectors, several conservation organizations. In the end, we agreed that the fishery met the MSC standard and it was certified. By 2017, six different toothfish fisheries, more than half the global catch, had received MSC certification. In response, MSC certified toothfish are no longer on the red list of the Seafood Watch program of the Monterey Bay Aquarium, the most commonly used seafood guide in the United States. For Greenpeace, though, the toothfish are still swimming on the red list.

Finally, on to the methuselahs of fishes, the orange roughy, who live to 150 years and are caught mostly in the deep waters of New Zealand and, in much smaller quantities, all around the world.

New Zealand orange roughy was a classic gold rush. A relatively small vessel could catch hundreds of thousands of dollars' worth of it in a single trip. No wonder the fishermen got a bit overenthusiastic and the stock was quickly overfished. At first, roughy graced the linen-tablecloth restaurants, but once science revealed just how badly depleted the fishery was, roughy quickly began to grace everyone's do-not-buy list. The quotas had been set much too high, and the New Zealand government had no choice but to reduce them and, in some cases, totally close certain areas to orange roughy fishing. And as hoped, as soon as fishing pressure let off, the stocks rebuilt. By 2016, the major roughy fisheries in New Zealand were certified by MSC. The stocks have rebuilt to the sustainable range, and the management system meets the MSC standard. Because of the gold rush character of poor early management and because some of the orange roughy fishing damages seamount habitat, no major environmental NGOs have taken orange roughy off the red list.

What Do We Really Know About the Global Status of Fish Stocks?

To begin with, there is no black and white in fisheries—only shades of gray. The outlook is generally good for fish stocks with scientific assessments and comprehensive management systems in place because abundance is mostly going up. Problem areas, like the Mediterranean, much of Africa, and some of Latin America, persist with either downward trends or poor overall stock abundance. But there are also areas with good average stock status that contain some overfished stocks.

Overfishing is very much in the mind of the beholder. In Table 5.1, we used the US government definition of *overfishing* that a stock is overfished when the abundance of the stock is less than half of the level that would produce long-term maximum sustainable yield. The concept of overfishing is very much tied to the objective of fisheries management. It implies that a stock is at a lower abundance than would be consistent with achieving management goals. Almost all national and international legislation is built around the concept of maximum sustainable yield—that is getting the best possible harvest over many generations. I like to look at it differently. Rather than worrying about whether a stock is called overfished, I suggest we look at how much long-term yield would be lost if we continue to manage fish stocks with the current level of fishing pressure. If we fish too hard, yield will be lost because of overfishing—but if we fish too little, yield will be lost because we are not taking advantage of the total potential harvest.

A Closer Look

Table 5.1 *The Fraction of Stocks Overfished and the Estimated Total Loss of Potential Yield from Excess Fishing Pressure*

	Percent of stocks overfished	Yield lost from fishing too hard
US Alaska	0%	0.0–0.0%
US West Coast	6%	0.0–0.0%
Pacific Ocean tuna	7%	0.5–1.2%
Indian Ocean tuna	11%	0.0–0.1%
South Africa	13%	0.4–1.0%
European Union Atlantic and Baltic	13%	2.0–4.9%
Australia	14%	8.5–11.6%
Canada West Coast	17%	0.5–1.0%
US Southeast and Gulf	18%	0.1–0.3%
Norway Iceland Faroe Islands	19%	2.4–5.3%
New Zealand	19%	1.2–2.4%
Atlantic Ocean tunas	29%	0.8–1.3%
US East Coast	32%	2.8–6.3%
Northwest Pacific	38%	1.2–3.6%
Canada East Coast	39%	2.1–3.6%
South America	43%	2.3–12.6%
Mediterranean-Black Sea	48%	20.4–48.0%
Total	23%	4.1–8.1%

The table shows that while many stocks (23 percent overall) would be called overfished, relatively little yield is being lost by excess fishing pressure (4.1–8.1 percent). Only in the Mediterranean-Black Sea is the loss from fishing too hard really very significant. How can a region such as Canada's East Coast have 39 percent of stocks overfished, but only be losing a very small amount of potential yield? This can happen for at least two reasons. The stocks may be at low abundance from too much fishing pressure in the past, as the infamous Northern cod, or the overfished stocks may be small and contribute little lost yield. In calculating the percent overfished, all stocks count the same, whereas in calculating lost yield, big stocks obviously count more than small stocks.

So what causes the differences between regions? It is fishing pressure. Where there are lots of overfished stocks, fishing pressure is or was too high, and where there is a trend for abundance to go up, fishing pressure has been reduced dramatically.

All this analyzing tells us that the so-common narrative about the global decline of fish is wrong for most of the majority of areas with good scientific data. But what about the other half of the world's fish catch? What happens there?

The Other Half of World Fisheries: Those That Are Not Scientifically Assessed

The picture in South and South East Asia, Africa outside of South Africa, and some of Latin America is not good. We do not have scientific assessment of trends in stock abundance, but we do have a wide range of other data that provide an insight. Perhaps most reliably, we have scientific surveys from some places.

For instance, the Gulf of Thailand has been routinely monitored by scientific surveys for decades. What we see there is severe depletion of the *demersal* (bottom-dwelling) and pelagic fish stocks, the mainstay of commercial fishing. Surveys, where they exist for bottom trawling, in South and South East Asia typically also show strong declines in abundance.

Fishing pressure in South East Asia is very high. Ordinarily, demersal fish are caught by trawling, dragging a net along the bottom. In many countries, we can measure fishing pressure by how many times a trawl net passes over the same point on the bottom of a country's continental shelf. Where stocks are healthy, a net on average passes over the same spot about once every 3 to 10 years. In South and South East Asia, that net passes 3 to 10 times per year. Small wonder that stocks are low.

We are lucky also to have expert opinion on the status of stocks in this region. My colleague Michael Melnychuk organized a study that asked people who are familiar with fisheries in various countries about the status of stocks in their region.[11] Uniformly, across South and South East Asia, most of Africa, and most of Latin America the answer was that stocks were in poor condition.

A Closer Look

Abundance

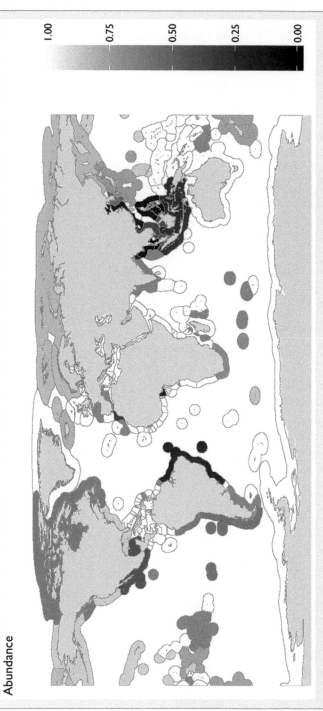

Figure 5.7 shows expert opinion on the status of fish stocks for twenty-eight of the most important fishing countries in the world. The dark shaded regions are the economic zones of countries where fish stocks are in poor shape. Of the countries surveyed, the health of stocks in South and Southeast Asian countries, as well as Brazil and Nigeria, were consistently rated as being poor.

The Asian Solution and Fishing Down Food Webs

Can We Really Say That All Those Unassessed Stocks Are Overfished and Declining? Not Really, Sometimes It Depends on Whom You Ask

Tim McClannahan, a participant in the NCEAS meetings, has worked on small-scale coastal fisheries in Kenya for almost 30 years for the Wildlife Conservation Society, formerly known as the Bronx Zoo. He is a marine ecologist who placed himself on one side of the divide between ecologists and fisheries scientists when he said at the first meeting, "where I work, the harder you fish the more you catch." Anathema had been pronounced. Obviously, for both ecologists and fisheries scientists, logic dictates that when you fish too hard, you overfish and you catch less.

Not so in coastal Kenya. Tim explained when you fish harder and harder in Kenyan coastal fisheries, the long-lived fish that eat the short-lived ones will soon be eliminated. That creates a much simpler ecosystem of a small number of highly productive, predator-free species, whose total yield (in weight) is as high, or maybe higher than it was under less fishing pressure. The downside is that the value of the catch declines. Big, long-lived predatory fish are worth more money per pound. Much of Tim's work with Kenyan fishermen is trying to convince them to fish less and make more money.

Another participant was Beth Fulton, a research scientist at the Australian Commonwealth Scientific and Industrial Research Organization, and arguably the world's expert on marine ecosystem modeling. Beth's reaction to Tim's story was, "oh, that is what I call the Asian solution." She meant that in much of Asia fishing pressure was so high that the long-lived predators were gone and the short-lived species happily kept on reproducing and boomed. Something similar happened when Europeans colonized North America. We shot and trapped the top predators, wolves, cougars, and bears, to make life safer for us and our livestock, and at the same time, released deer from predation and turned them into lawn and garden nuisances.

Recently, Cody Szuwalski and co-authors looked at the Asian solution in Chinese fisheries. They agreed with McClannahan that the removal of predators had led to a great expansion of their prey species, what ecologists call a *trophic cascade*. In this case, they felt that if the Chinese were to apply the single-species management used in the United States and other countries, the results would not at all be for the better.

> Single-species management would decrease both catches and revenue by reversing the trophic cascades. Our results suggest that implementing single-species management in currently lightly managed and highly exploited multispecies fisheries (which account for a large fraction of global fish catch) may result in decreases in global catch. Efforts to reform management in these fisheries will need to consider system wide impacts of changes in management, rather than focusing only on individual species.[12]

What Szuwalski and his collaborators recommend is what is commonly called *ecosystem-based fisheries management*, but not along the US model that usually includes cautious single-species management. If the Chinese were to rebuild their predatory species to the level that would provide maximum sustainable yield as is mandated by US fisheries law, they would dramatically decrease both the tonnage and value of their catch.

This is a revolutionary suggestion to most marine conservation groups for whom rebuilding all species is a core tenet of their worldview.

It is no revolution for Beth Fulton and the small group of specialists who work on marine ecosystem models. Here we need to digress into some simple ecosystem ecology. We classify marine species by their trophic level: where they are in the food chain. Trophic level 1 is the plants that convert sunlight into carbohydrates. Level 2, zooplankton like krill, eats level 1. Fish generally begin at trophic level 3 and eat level 2. The trophic level of any species is defined by what it eats.

The amount of energy that flows up the food chain is reduced 90 percent at each trophic level. If we want to maximize total energy production, we would harvest as low down on the food chain as possible, which is what we do on land—we mostly eat plants, trophic level 1, and herbivorous livestock, trophic level 2. In contrast, in marine ecosystems, we primarily eat trophic levels 3 to 5, although some invertebrates such as mussels and clams belong to levels 2 to 3. What McClannahan, Szuwalski, and Fulton have shown is that, if we want to get the most food from marine ecosystems, we should harvest low on the food chain.

This has led to one of the more interesting debates about fisheries management in the last decade, called *balanced harvesting*. At present, we are picky, and we like to catch and eat higher-trophic-level species, and the bigger the better. The proponents of balanced harvesting will tell you that it . . . *distributes a moderate mortality from fishing across the widest possible range of species, stocks, and sizes in an ecosystem.* Beth Fulton calculated that non-selective fishing can produce perhaps 30 percent more yield.

By now you should not be surprised that balanced harvesting was raked under the harrow of criticism by other marine ecosystem ecologists concluding that . . . *this body of evidence suggests that BH (balanced harvesting) will not help but will hinder the policy changes needed for the rebuilding of ecosystems, healthy fish populations, and sustainable fisheries.*

We forge on to the next controversy. This one is about fishing down food webs. In 1998, Daniel Pauly of the University of British Columbia and his colleagues described it as *a gradual transition in landings from long-lived, high trophic level, piscivorous bottom fish toward short-lived, low trophic level invertebrates and planktivorous pelagic fish.*[13] It is one of the most famous papers in fisheries ever published and is referred to by other scientists over 4,500 times. The underlying premise is simple: large-bodied valuable fish like bluefin tuna tend to be long lived and at a high trophic level and are the initial targets of developing fisheries. When they are depleted, fishing switches to the less valuable but more abundant species that are lower on the food web until, in the end, we will have jellyfish sandwiches for lunch.

Pauly and co-authors supported this argument by showing that the mean trophic level of fish in the FAO landings database had declined.

Gentle Reader, Do You Hear the Snap of the Method Trap?

At first, I never doubted that it was true but I did wonder if it might possibly not be bad. After all, we can produce more food from the ocean by fishing lower on the food chain.

Back to NCEAS once more. Trevor Branch, then a post-doctorate at the University of Washington, took a leading role in the development of the survey database we used in our 2009 paper. He wanted to see if the mean trophic level of the ecosystems was declining in the survey data and he linked the survey abundance data with a database known as FishBase for trophic levels for each species. To everyone's surprise he not only found that the mean trophic level in surveys did not decline, he also found that neither did the mean trophic level of the landings in the FAO data—it had actually increased in recent decades.[14] The paper that received 4,500 citations from other scientists was simply wrong. In other words, we could replace the jellyfish sandwich with an equally plausible bluefin tuna steak. Even though in some ecosystems the mean trophic level is indeed declining, globally, Pauly's trend is simply not there.

Then, to hammer some more nails into the coffin of fishing down marine food webs, Suresh Sethi, a PhD student in my lab, looked at the idea that fishing down is caused by fishermen first targeting the large valuable fish. He linked the trophic-level database to a price-per-pound database maintained by Pauly's The Sea Around Us project and showed that there was absolutely no relationship between price per pound and trophic level.[15] Some of the most valuable species, abalone, scallops, and shrimp, are low down on the food chain, and many high-trophic-level fish like skipjack tuna and sharks are not particularly valuable. Sethi then looked at the history of how fisheries develop over time and what species are caught first. The answer turned out to be pretty simple: abundant fish stocks tend to be targeted first. Fishermen go where the money is.

It is truly difficult to combat the perception that the oceans are being emptied of fish. But there is hope. Villy Christensen, who, like Beth Fulton, is one of the world's experts on marine ecosystem models, used more than 200 such models, for a study he published in 2014, to estimate how the number of fish in the sea had changed.

> Our results predicted that the biomass of predatory fish in the world oceans has declined by two thirds over the last 100 years. . . . Results also showed that the biomass of prey fish has increased over the last 100 years, likely as a consequence of predation release.[16]

There it was, the most optimistic assessment of global fisheries ever. We know from basic population dynamics that to get the best possible harvest over the long term, we must bring down fish populations to roughly 30 to 40 percent of their original abundance. Using Christensen's calculations, on average, we have reached that target for the predatory fish. But if the lower- trophic-level fish have truly increased and are much, much more abundant than their predators are, then, maybe instead of emptying the oceans of fish, there are actually more tonnages of fish in the ocean now than before industrial fishing began. Now that is a truly revolutionary thought.

To follow this up, I got in touch with Jeppe Kolding, who also works with ecosystem models. He intrigued me the first time we met by stating that Africans using mosquito nets to catch tiny fish was a fine idea. This, of course, is the very opposite of the common narrative that fishing with very fine nets is destructive, and repurposing nets provided by US foundations to protect people from malaria in order to plunder the lakes of Africa is doubly pernicious.

Kolding had been a nurse in hospitals in East Africa. He said that once you have seen children die from malnutrition, you look at the world differently than an ecologist sitting in a University in Denmark or the United States. Small fish are one of the most nutritious foods known, and these very small fish in the African Great Lakes are not overfished. And if they can be caught with mosquito nets, all the better.

On my prodding, Kolding analyzed all the ecosystem models he had and concluded that indeed there were as many or more tons of fish now than there had been before industrial fishing began for temperate fisheries like the United States and Europe that target high-trophic-level fish. In the tropics, on the other hand, where harvest is much less selective, there were undoubtedly fewer fish now.

I hope that, with this chapter, I was successful to convince you, the skeptical reader, that the oceans are not being drained of fish and that in most places where we have good scientific data, fish stocks are doing well and are increasing. For the other half of the world, we may have to readjust our goal. Even though there is convincing evidence that, particularly, the long-lived species are severely depleted in many areas, this altered state does appear to be able to produce the near maximum catches to provide food for people.

· ·

FURTHER READING

An early paper warning of the dangers of overfishing. Ludwig D, Hilborn R, and Walters C. 1993. Uncertainty, resource exploitation, and conservation—lessons from history. Science 260: 17.

The paper that suggested predatory fish have declined by 90 percent. Myers R A and Worm B. 2003. Rapid worldwide depletion of predatory fish communities. Nature 423: 280–3.

The paper criticizing the estimated 90 percent decline of predatory fish. Hampton J, Sibert J R, Kleiber P, et al. 2005. Decline of pacific tuna populations exaggerated? Nature 434: E1–E2.

The classic history of cod. Kurlansky M. 1997. Cod: A biography of the fish that changed the world. New York: Walker and Co.

The original paper projecting all stocks would be collapsed by 2048. Worm B, Barbier EB, Beaumont N, et al. 2006. Impacts of biodiversity loss on ocean ecosystem services. Science 314: 787–90.

The paper that showed "all fish gone by 2048" was wrong. Worm B, Hilborn R, Baum JK, et al. 2009. Rebuilding global fisheries. Science 325: 578–85.

A paper showing why using catch data is the wrong way to estimate stock status. Branch TA, Jensen OP, Ricard D, et al. 2011. Contrasting global trends in marine fishery status obtained from catches and from stock assessments. Conserv Biol 25: 777–86.

The paper that debunked the global fishing down food chains. Branch TA, Watson R, Fulton EA, et al. 2010. The trophic fingerprint of marine fisheries. Nature 468: 431–5.

The paper estimating that predatory fish had declined by 60 percent, but lower-trophic-level fish had increased in abundance. Christensen V, Coll M, Piroddi C, et al. 2014. A century of fish biomass decline in the ocean. Mar Ecol Prog Ser 512: 155–66.

The original fishing-down-marine-food-webs paper. Pauly D, Christensen V, Dlasgaard J, et al. 1998. Fishing down marine food webs. Science 279: 860–3.

Introduction to the concept of balanced harvesting. Law R, Plank MJ, and Kolding J. 2012. On balanced exploitation of marine ecosystems: Results from dynamic size spectra. ICES J Mar Sci 69: 602–14.

The final nail in the coffin of fishing down marine food webs. Sethi SA, Branch TA, and Watson R. 2010. Global fishery development patterns are driven by profit but not trophic level. Proc Natl Acad Sci USA 107: 12163–7.

A description of the RAM Legacy Stock Assessment Database. Ricard D, Minto C, Jensen OP, et al. 2012. Examining the knowledge base and status of commercially exploited marine species with the RAM Legacy Stock Assessment Database. Fish Fish 13: 380–98.

A reassessment of the status of Chinese fisheries (and thus most intense tropical fisheries). Szuwalski CS, Burgess MG, Costello C, et al. 2016. High fishery catches through trophic cascades in China. Proceedings of the National Academy of Sciences: 201612722.

CHAPTER 6

The Environmental Impacts
of Fishing

"As an environmentalist, should I stop eating fish?" Markus Borner asked, waving the cover of the November 2009 issue of *Time* magazine at me that showed chopsticks holding a piece of tuna sashimi, dripping blood. All the tuna in the world were disappearing, the article said, and Markus wanted to know if he, as a person who had devoted his life to conservation, should give up eating the fish he loved.

A chance question like that can change the course of one's career. It happened in 2010, in Serengeti National Park, in Tanzania, when I helped the Frankfurt Zoological Society estimate the size of the park's famous migrating wildebeest herd. Markus Borner, who was then the head of the Frankfurt Zoological Society's programs in Africa, and I sat on the verandah enjoying the sunset and drinking beer, when he popped the question.

I countered with "if you don't eat fish what will you eat instead?" Markus, a confirmed carnivore, took me out to the road and said:

> If you go down that road 150 kilometers you will get to Olduvai Gorge. A million years ago my ancestors came down from the trees and ever since then have been working their way up the food chain. I couldn't possibly dishonor their memory by being a vegetarian. If I stop eating fish, I will eat more beef, chicken and pork.

Markus's question nagged and led me to try to understand how the food we eat impacts the environment. On return to Seattle, I proposed to my department that I teach a new course for first-year students on food and the environment, and I have been teaching it ever since. With time, my fisheries focus shifted from primarily stock assessment to fisheries as a source of food and food security.

I am no stranger to food production, albeit mostly terrestrial, such as harvesting peas in Eastern Washington for Jolly Green Giant as a student. Thirty years later as a full professor, I became a farm hand once again, working for my son, who, somewhat inexplicably, became an industrial farmer on Whidbey Island, 60 miles north

Ocean Recovery: a sustainable future for global fisheries? Ray Hilborn and Ulrike Hilborn, Oxford University Press (2019). © Ray Hilborn and Ulrike Hilborn 2019. DOI: 10.1093/oso/9780198839767.001.0001

of Seattle in Puget Sound. I worked on weekends for my wife, who, equally unexpectedly, started growing 120 varieties of organic vegetables. Despite hands-on farming, I had to make the leap from farmhand on a tractor to understanding the academic literature on the environmental impacts of producing food.

Leading the way, Peter Tyedmers's seminal work on fuel use by fishing fleets[1] had already suggested that, particularly when measuring greenhouse gas output, many forms of fishing were gentler on the environment than producing Markus's alternatives of beef, chicken, and pork. The next big find was a 2006 FAO report entitled "Livestock's Long Shadow: Environmental Issues and Options,"[2] which should have put anyone off eating livestock with its gory details regarding desertification, water consumption, manure pollution, carbon footprint, and soil erosion. Then finally, a broad set of hard data appeared. Nathan Pelletier's paper on "Energy Intensity of Agriculture and Food Systems"[3] is full of graphs on energy use by quite a range of livestock, capture fisheries, crops, and aquaculture. On average, Pelletier says, capture fisheries use less energy than beef and most aquaculture do, and roughly the same as poultry, pork, and lamb. But if you were selective about the kind of fish you eat, such as sardines, herring, or anchovies or aquaculture species like mussels and oysters, you would lower energy used to be comparable to a vegetarian diet.

How Does One Scale Up the Energy Used to Catch a Fish or Raise a Pig into Numbers that Allow Comparison?

We have Coca Cola to thank for the method that calculates what goes into a product and what comes out. In 1969, they looked at different containers for their brown liquid and quantified the raw materials used and the environmental outputs. The method is forever being refined, and as life-cycle assessment (LCA), it is routinely used for any industrial input-output study. There are books and journals, even a hitchhiker's guide to LCA, and should you want to build a ship, an LCA software package will tell you how much energy will be used to manufacture the steel you are planning on buying.

Let us assume you live in Europe and are hoping for a rack of lamb for dinner. You ponder your choices laid out for you in the meat department of your local supermarket. There is local lamb, nice and fresh, and there is vacuum-packed New Zealand rack of lamb, from the very other side of the world. You are not too worried about the price and have been supporting local farmers and producers, so you reach for the European lamb. But wait—what about the carbon footprint? Now the case is altered. With a cradle-to-grave, well, let us call it cradle-to-dinner-table LCA, the New Zealanders have shown that despite raising their lambs down under, they are so efficient at it on the farm and in shipping that only 5 percent of total greenhouse gas release comes from transport. Europeans farm much more intensively and use more fertilizer, giving the antipodean rack of lamb a smaller carbon footprint than the European one.

The New Zealand cradle-to-dinner-table LCA is an exception to the standard for food LCAs that are normally calculated from cradle to farm gate or fish arriving at the dock. It is Peter Tyedmers and his students and post-doctorates who have set the fisheries LCA standards and to whom both industry and NGOs turn for the numbers, as did I, reviewing their papers in search for a better answer to Markus Borner's question, and with that, I entered the world of food life- cycle assessment.

Roughly half of the fish we eat at present come from aquaculture, which is rather more like farming than catching wild fish. Little fish are raised in a hatchery and then "grown out," commonly in ponds for shrimp, catfish, and tilapia, but open net pens in the ocean for salmon. Shellfish like mussels and oysters are planted out in the wild on rafts or wire cages.

Getting in touch with Steve Hall brought me a veritable avalanche of LCAs for aquaculture. Hall was then the Director of the World Fish Center in Panang, Malaysia, that describes itself as "... an international, nonprofit research organization that harnesses the potential of fisheries and aquaculture to strengthen livelihoods and improve food and nutrition security" with a focus on tropical countries, particularly in Asia. By happy coincidence, he was in the midst of life-cycle assessments for aquaculture, and fifty-one of them culminated in the report "Blue Frontiers: Managing the Environmental Costs of Aquaculture."[4] Moreover, he and his colleagues had looked at direct comparisons of aquaculture to raising livestock for kilograms of nitrogen released per ton of protein produced. Beef clocked in at 1,200 kilograms; pork at 800 kilograms; and chicken at 300 kg. Aquaculture fish averaged 360 kilograms, whereas farmed shellfish trumped everything else at -27 kilograms since they clean the water by taking up nitrogen rather than releasing it. To put that staggering number for beef in perspective, 1,000 kilograms of nitrogen equal 760 cubic meters of manure. That is a really big hill.

Since 2010, several students, research assistants, and I have found 381 LCAs of food systems. These assessments calculate a wide range of environmental impacts, but we have focused on greenhouse gas, energy used, acidifying compounds released, eutrophication compounds released, area of land used, and amount of water used.

What does a ton of beef look like? Or a ton of milk, which is mostly water? A ton of anything is not something that comes to mind easily, so grappling with the enormity of the ton, I downsized the unit of measurement for any comparisons to 40 grams of protein, that is what is found in a typical 8 ounce serving of fish or meat, and all the daily protein you need to sustain yourself. And yes, that 32 ounce steak really is overkill.

Meanwhile, Let Us Tease Out the Meaning and What We Can Learn from the Many Assessments

The winners for lowest emission of greenhouse gases, generated by fertilizer for plants, fuel for capture fisheries, and feed for farmed salmon, are aquaculture for

A Closer Look

This table shows a range of environmental impacts calculated per 40 grams of protein produced. These are the average of available live-cycle assessments.

	Antibiotics (mg)	Nutrient release (kg PO4)	Fuel use (MJ)	GHG (kg C02)	Land (square meters)	Water (cubic meters)
Aquaculture						
Salmonids	113	0.02	17.06	1.00	0.79	13.50
Shellfish	NA	−0.01	5.07	0.31	1.88	NA
Tilapia	NA	0.05	42.79	3.60	4.64	68.90
Invert	NA	0.03	44.94	4.83	3.70	2.60
Whitefish	NA	0.06	58.04	4.84	8.57	NA
Livestock						
Lamb	NA	NA	3.09	3.47	64.75	0.70
Beef	54	0.12	11.55	6.43	21.91	0.93
Chicken	124	0.01	6.03	1.57	2.62	0.57
Dairy	NA	0.02	3.41	1.34	1.70	0.47
Pork	174	0.07	7.58	1.70	3.01	1.23
Capture Fisheries						
Invert	0	0.02	56.42	6.76	17.00	0.00
Large pelagic	0	0.00	9.07	2.48	NA	NA
Shellfish	0	0.00	3.13	0.04	NA	NA
Small pelagic	0	0.00	3.15	0.72	6.05	0.31
Whitefish	0	0.00	13.49	2.23	27.13	0.01
Plants						
Corn	0	NA	4.20	0.58	NA	NA
Potatoes	0	0.00	3.36	0.38	0.72	0.05
Rice	0	0.01	1.88	0.80	NA	0.23
Soy	0	NA	0.71	0.08	NA	NA
Wheat	0	NA	0.95	0.15	NA	NA

shellfish and salmon, all plants, particularly soy and wheat, and capture fisheries for small pelagic and whitefish.

The picture changes for energy use. Shellfish came in lowest in aquaculture; livestock, excepting beef, is reasonably similar; and small pelagic and shellfish fisheries remained competitive. All plants do well in this category.

When measuring nutrients released, almost all capture fisheries tended to look very good; milk and chicken had low impact; and shellfish aquaculture was the clear overall winner.

Water—without it we are nothing, we do not exist. Yet, in those parts of the world where it rains, we use it with great abandon. When next you down your glass of almond milk, consider that it takes 1.1 gallons of water to grow one, repeat, *one* single almond. Where water is limited, agriculture is equally limited. Where the desert encroaches, agriculture stops. But what about capture fisheries? The LCAs show us that fishing boats use very little, if any freshwater. If boats carry ice, that counts as freshwater. Some water is used in building a fishing boat, and almost all boats carry some for the crew. Compare this use of precious little freshwater by capture fishing boats to that of livestock. From the limited number of LCAs available, it takes close to 1 cubic meter, that is 1,000 liters or 264 gallons, of water to produce 40 grams of livestock protein, whereas capture fisheries use almost none. Aquaculture water use is muddied—how does one measure water used for species reared in ponds? There are, of course, finer points to consider, such as is the water used for raising livestock and crops from irrigation only or does it come from rainfall or both. But from the perspective of Markus's question, the answer is very clear, capture fisheries win.

For the area used to produce our 40 g of protein, there are pretty good data in square meters. Lamb and beef are high because much is raised by grazing; chicken and pork are much lower because they are fed crops that have high yield per area; capture fisheries are difficult to evaluate because they do not require land conversion to produce; but most estimates end up with high numbers because they simply look at the yield of fish per area of the ocean.

When it comes to antibiotic use, capture fisheries, farmed shellfish and organic livestock win hands down. They use none. Most aquaculture for fisheries and livestock use some. There is legitimate growing concern that pervasive preventative antibiotic use in standard animal production is breeding antibiotic resistant bacteria, such as flesh eating bacteria, and, therefore, it must be considered one of the environmental consequences of food production.

Pesticides and fungicides are generally used in making feed for most livestock and aquaculture. Here organic agriculture does not get a free pass, since using naturally occurring pesticides such as pyrethemum and rotenone is permitted. Capture fisheries do use some toxic compounds in the antifouling paints put on ships hulls, but the amounts are small and mostly copper based. Because small amounts of copper are beneficial to phytoplankton in marine waters, antifouling paint's impact is neutral.

Finally, we come to soil. The finest guide to understanding soil is *Dirt: The Erosion of Civilizations*[5] by David Montgomery, one of my colleagues at the University of Washington, a geologist whose interests range from salmon restoration to landslides to food production. Dirt sustains human society, and we know that the loss of topsoil has brought down many a civilization. I use Montgomery's online lectures in my food course with the central theme that the biggest threat to long-term food sustainability is not so much climate change but the loss of topsoil. As soon as we invented the plow, topsoil began to vanish as erosion increased 100 fold. In the United States, even though the plow has had a reasonably short history, 30 percent of US topsoil has eroded away since its introduction. Globally, given the current rates of soil loss, the capacity to produce food will be greatly diminished within a century.

> Man—despite his artistic pretensions, his sophistication, and his many accomplishments—owes his existence to a six-inch layer of topsoil and the fact that it rains.
>
> - Author unknown

Most of our food production systems, crops, livestock, and aquaculture of species that are fed plant matter contribute to the decline of our topsoil, whereas farmed shellfish, algae that are not fed, and capture fisheries do not diminish our dirt.

I do not usually have the clever answer to a question or a brilliant comment during a conversation, but rather it occurs to me later, often years later. I certainly failed to provide Markus Borner with the right answer at the time. What I should have responded to Markus, who spent his career trying to preserve and enhance the biodiversity of Africa, was this: To produce beef, chicken, and pork, the alternative to eating fish, you need to grow crops. Growing crops means, in almost all cases, eliminating a natural ecosystem and its vegetation and replacing it with highly productive exotic species. Capture fisheries do not impact the phytoplankton, the marine equivalent of terrestrial plants, and they very rarely impact the small animals who eat it. What capture fisheries do is reduce the abundance of targeted species, but they do not transform, that is destroy, natural ecosystems as agriculture does. The biggest threat to wildlife in Tanzania has not been poaching (which is certainly serious for rhinos and elephants), but the pressure for more farmland. My friend Tony Sinclair has been recording biodiversity both inside and outside of Serengeti for decades, and as soon as you reach farmland, biodiversity plummets. If Markus were to stop eating fish in favor of beef, chicken, or pork, he would, in a small way, contribute to the steady trend of the last 50 years to convert African forests and savannah to farmland. Had I been quick enough to think of this when Markus asked his question, we might have settled the issue right then.

My wife used to grow organic vegetables on 5 acres of land without synthetic fertilizers or pesticides and, presumably, low biodiversity impact. On the face of it, a laudable enterprise. However, 150 years ago, her 5 acres were covered by temperate rainforest. There were Douglas fir, hemlock, and cedar trees, an understory of nurse logs, salal, ferns, and berries and its unique community of birds, mammals, reptiles, amphibians, insects, and other invertebrates—now all gone.

But if you look at the waters of Puget Sound, 300 yards from my wife's 5 acres, the species that were there 150 years ago are there still. Certainly some have been overfished and are at low abundance, yet the marine ecosystem today is close to what it was 150 years ago. Moreover, if Puget Sound had been well managed and overfishing prevented, it would now take a detailed scientific study to see any difference.

A Closer Look

Figure 6.1 shows the changes that high seas tuna fishing has caused in the ecosystem structure—the fisheries that Markus Borner was worried about because that is where his tuna sushi comes from. We see some change of abundance at the top of the food chain where the tuna are—but only about 20 percent change.[6]

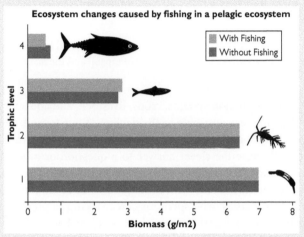

Figure 6.1 Changes in trophic structure of a pelagic ecosystem. Data from Tim Essington.

In contrast, Figure 6.2 shows the change in abundance of different species and groups outside Serengeti Park compared to inside. Everything but rodents outside the park is almost gone.

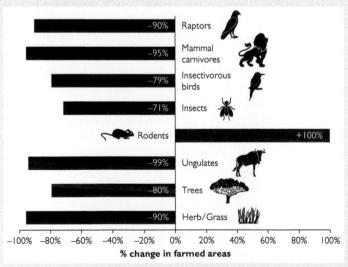

Figure 6.2 Changes in biodiversity outside Serengeti National Park compared with Inside. Data from A.R.E. Sinclair.

Of course, some fishing practices have been and continue to be nearly as transformative as farming. Dragging heavy nets over corals and sponges can destroy the existing ecosystem that can take as long to recover as an old growth forest. There is no question that this should be stopped and the sensitive habitats protected. But most of the world's fisheries take place in ecosystems that are remarkably little changed by fishing except for the reduction in abundance of the target species.

Biodiversity: Menhaden versus Soy Beans

In my attempt to compare the biodiversity impacts of capture fishing to those of farming, I find it very difficult to turn up numbers for the biodiversity loss from farming. However, bear with me as we look at the Atlantic menhaden fishery and what it would mean were we to replace it with food from the land.

In 2017, management of the Atlantic menhaden on the East Coast of the United States was a hot topic. Menhaden are abundant, herring-like fish that are turned into fishmeal and oil that is fed primarily to aquaculture-reared Atlantic salmon. Environmental groups advocated reducing the commercial harvest to make more menhaden available to their predators, a wide range of marine fish and some marine mammals, although there is no evidence that either predatory fish or marine mammals on the East Coast are going hungry.

Nevertheless, let us assume we reduce the menhaden harvest by 100,000 tons, that is roughly half the present harvest, and replace the lost protein by growing soybeans in Brazil. This is a reasonable assumption because salmon farmers are more and more turning to soybeans to meet the increasing demand for protein to feed salmon.

Now to the numbers. To produce the digestible protein equivalent of 100,000 tons of menhaden we need 157,000 tons of soybeans. To grow this many soybeans will take 62,000 hectares or 621 square kilometers of arable land. To clear 621 square kilometers of Brazilian rainforest, 22 million tons of natural vegetation will go up in smoke, releasing 81 million tons of CO_2, the equivalent to the CO_2 emissions of 17 million cars. At the same time, 80,000 parrots, 13,000 insectivorous birds, twenty-seven jaguars, sixteen maned wolves, and billions of insects and other mammals, reptiles, and amphibians will suddenly be homeless.

It would seem that using the precautionary approach to harvest fewer menhaden will have a catastrophic impact on a terrestrial ecosystem far away from our eyes.

Okay, but What About Extinction?

Extinction is perhaps the most misunderstood aspect of marine fish population. If nothing else, it makes good headlines. Thus, the headlines *Overfishing Driving Dozens of Species into Extinction*, [*The Independent*], *Salt-Water Fish Extinction Seen by 2048*, [CBS News Web Site] and *Ocean life faces mass extinction* [*The New York Times*]. And so on.

The public has certainly been persuaded. Take New Zealand, a country where at least 190 endemic bird species are known to have gone extinct since the arrival of humans who came with rats, pigs, dogs, and cats that munched their way through massive populations of birds that had lived happily without terrestrial predators. Most of the arable and grazeable land has been converted from native forest to exotic pine plantations for lumber, fruit orchards, vineyards, corn fields, and pastures for dairy cows and sheep; 90 percent of New Zealand's wetlands have been drained primarily for agriculture. Yet, in a Lincoln University survey of residents, New Zealanders rated their country as "clean and green" with air, native bush, and forests in their best condition.

Rivers, lakes, and marine fisheries, on the other hand, were thought to be in the worst state, even though no marine fish species there has ever gone extinct.

Elsewhere in the world, the only marine fish anywhere to go extinct was the Galapagos Damselfish, a species that was never fished and disappeared after 1983 because of habitat loss and climate change.

About twenty marine species that have indeed gone extinct have been marine birds and mammals that were easily caught and hunted into oblivion. A few marine plants and invertebrates have also made their permanent exit, and next on the potential list of extinction is the vaquita, a small porpoise that lives only in the northern Gulf of California and has a population of fewer than one hundred.

What If I Become a Vegetarian?

Many people choose to be vegetarians or vegans believing that they are preventing animals from suffering and being killed. Yet there is a 2011 article by Mike Archer of the University of New South Wales with the provocative title, "Ordering the Vegetarian Meal? There's More Animal Blood On Your Hands," where he claims:

> If you want to minimise animal suffering and promote more sustainable agriculture, adopting a vegetarian diet might be the worst possible thing you could do...Published figures suggest that, in Australia, producing wheat and other grains results in at least 25 times more sentient animals being killed per kilogram of useable protein, more environmental damage, and a great deal more animal cruelty than does farming red meat.

Hold on a Minute, How Does That Work?

Our son farms on an idyllic spot on Whidbey Island in Washington State. There is no shortage of bald eagles on the island, and when it is time to cut grass and corn, the eagles are on high alert, and they congregate in trees and on telephone poles by the dozen to see what the approaching chopper has to offer, while the bird watchers with long lenses cause traffic jams on the road. Mice, rats, and rabbits flee in all directions; some will get chopped up along with the occasional fawn, too well hidden for the man on the tractor to see it, and the eagles swoop in. Whether dead or alive, it is all a free lunch for them. Now extrapolate that to the infinite variety of small and middling marsupials that hang out in Australian fields, and you will see Archer's point.

In defense of vegetarians, it must be said that grass and corn for silage are raised to feed heifers that are raised to produce milk. Although, when I harvested peas for the Jolly Green Giant as an undergraduate, the same thing happened. Wherever I drove the combine, rabbits and other small mammals shot out from the rows and a predator picked them off. Those who did not run were combined with the peas— not a pleasant image for a vegetarian, I have to admit.

Everyone has to eat and eating invariably causes some kind of death. If we accept that terrestrial food production drastically reduces native habitat, we must also spare a thought for the lives that never happened because their habitat was gone.

In sum, I hope to have made the point that, in general, eating fish that are sustainably harvested from the sea has a lower environmental impact than the alternatives of livestock or even a vegetarian diet. Therefore, well-meaning "precautionary" reductions in fishing pressure will have inevitable consequences that lost protein from the ocean will need to be replaced by protein from the land. If not, famine follows.

• •

REFERENCES AND FURTHER READING

Tyedmers early work on fuel use in fisheries. Tyedmers P. 2004. Fisheries and energy use. Encyclopedia of Energy 2: 683–93.

An overview of energy use across food systems. Pelletier N, Audsley E, Brodt S, Garnett T, Henriksson P, Kendall A, Kramer K J, et al. 2011. Energy intensity of agriculture and food systems. Annual Review of Environment and Resources 36: 223–46.

The most recent synthesis of fuel use and green house gas emissions in fisheries. Parker R W, Blanchard J L, Gardner C, Green B S, Hartmann K, Tyedmers P H, and Watson R A. 2018. Fuel use and greenhouse gas emissions of world fisheries. Nature Climate Change 8: 333.

A major report on the environmental cost of livestock production. Steinfeld H, Gerber P, Wassenaar T, Castel V, Rosales M, and de Haan C. 2006. Livestock's long shadow: Environmental issues and options. Rome, Italy: Food and Agriculture Organization of the United Nations, 390.

LCAs of a range of aquaculture production systems. Hall S J, Delaporte A, Phillips M J, Beveridge M, and O'Keefe M. 2011. Blue frontiers: Managing the environmental cost of aquaculture. Penang, Malaysia: The World Fish Center.

Our group's paper comparing a range of impacts across animal source protein. Hilborn R, Banobi J, Hall S J, Puclowski T, Walsworth T E. 2018. The environmental cost of animal source foods. Frontiers in Ecology and the Environment16(6): 329–35.

The source of the data on changes in ecosystem structure caused by fishing. Essington T E. 2006. Pelagic ecosystem response to a century of commercial fishing and whaling. Whales, whaling, and ocean ecosystems. 1st edn. Berkeley: University of California Press, 38–49.

Mike Archer's entertaining look at how animals die in agriculture. Archer M. 2011. Ordering the vegetarian meal? There is more blood on your hands. The Conversation 16: https://

theconversation.com/ordering-the-vegetarian-meal-theres-more-animal-blood
-on-your-hands-4659.

A look at marine extinctions. Dulvy N K, Pinnegar J K, and Reynolds J D. 2009. Holocene extinctions in the sea. In: S T Turbey (ed.), Holocene extinctions. Oxford: Oxford University Press, 129–50.

Surveys showing New Zealanders think that fish in the ocean are more threatened than animals on land. Hughey K F, Kerr G N, and Cullen R. 2013. Public perceptions of New Zealand's environment. Lincoln, New Zealand: Lincoln University Press.

All you need to know about soils. Montgomery D R. 2007. Dirt: The erosion of civilizations. Berkeley: University of California Press.

CHAPTER 7

Recreational Fishing

One hand holds the hook; the other hand holds a worm; the worm ends up on the hook that is attached to a line that is attached to a pole; hook and bait hit the water; and there is a happy frisson of anticipation that a fish will be caught.

Be it from the banks of a stream, the end of a pier, or on board a thousand-dollar-a-day charter boat, the pleasure of recreational fishing is shared by anywhere from 200 to 700 million people. There are numbers for developed countries that are more reliable where 10 percent of the population, that would be 118 million people, fish for pleasure.

In the developing world, it is difficult to parse out recreational fishing from small-scale subsistence fishing, but it is clear that the more developed a country is, the more people go fishing for recreation. There is even an FAO definition for it: ". . . fishing of aquatic animals (mainly fish) that do not constitute the individual's primary resource to meet basic nutritional needs and are not generally sold or otherwise traded on export, domestic or black markets."[1] This distinction gets a bit murky because some trading of recreationally caught fish is quite common, and some jurisdictions even allow recreationally caught fish to be sold, blurring the line between small-scale subsistence and recreational fishing even further.

There is much diversity in where to fish and how to fish. From a simple hook-and-line to high-tech charter fishing, people also fish with traps, nets, and longlines. They fish in freshwater streams and lakes, in the neighborhood pond, on ocean beaches and out at sea. They fish on private land where owners charge a fee and on public land with nearly unrestricted use. Even though the recreational harvest on the ocean is small compared with industrial fishing fleets, for some species, recreational fishing surpasses the industrial catch.

Ocean Recovery: a sustainable future for global fisheries? Ray Hilborn and Ulrike Hilborn, Oxford University Press (2019). © Ray Hilborn and Ulrike Hilborn 2019. DOI: 10.1093/oso/9780198839767.001.0001

One Thing Is Certain: Recreational Fishing Is Very Big Business

On an evening in late August, while waiting for our luggage in the Seattle airport, the floor between two luggage carousels was covered in fish boxes that had all come from one resort on a previous flight from Sitka, Alaska. A harried guide tried sorting the boxes by owner, each box marked either halibut, salmon, or rock fish, while some anxious owners hovered to make his job a little harder. Let us say there were nearly thirty boxes of equal size, each box contained roughly 50 pounds of fish, each one of the fishermen (they were all men) brought home three boxes, that is 150 pounds of fish. In all, we were admiring, not without a tiny touch of envy, roughly 1,500 pounds of recreationally caught Alaskan wild fish.

Now, Let Us Get Into the Really Big Numbers

The European Commission estimates that marine recreational fisheries in the EU have a total economic impact of 10.5 billion Euros and support 100,000 jobs. The commercial fishery in 2014 generated an estimated income of 7 billion Euros and employed 109,000 people. On the other side of the globe, in 2011, Americans spent $48 billion in retail sales for recreational fishing activities, whereas the landed value of US commercial fisheries in that year was $5.3 billion. What does that tell us? In developed countries, recreational fisheries are more or less on par economically with the commercial fisheries, but when we consider the ancillary economies, that knock-on effect in services such as charter boat operators, and employment in equipment shops and resorts, and, particularly, the sheer number of people who participate, recreational fisheries weigh in rather heavily.

Fishing to What Purpose: Food or Pleasure?

The purpose of commercial fisheries is to produce food and make money, best achieved by highly efficient fleets with the lowest amount of effort.

Recreational fishermen want to fish for fishing's sake. This can best be achieved with highly inefficient gear and lots of effort. A day on a trout stream with a bag of only one trout is vastly preferable to a day at the office with a traffic jam.

However, a dead fish is a dead fish whether recreational or commercial. And that is where management comes in.

Gear restrictions, size limits, closed seasons, and closed areas are the most common tools used to manage both fisheries. Many recreational fisheries use daily or season bag limits, that is how many fish a person can kill, somewhat like trip limits that can be imposed on commercial vessels. Catch and release, on the other hand, is unique to recreational fishing. To let go what you have caught makes no sense in commercial fishing, but if the fish you release is likely to survive and you or your friends can catch it again—well, that makes that recreational fishery even more valuable.

In freshwater, there are generally more fishermen than fish, and to even out this discrepancy, for some species like trout, hatcheries that produce fish for stocking have sprung up like mushrooms. In a put-and-take fishery, a public or private agency puts hatchery fish into a lake or stream, and the recreational fishery does its very best to take them out again. In the United States, the biggest fisheries are for bass, panfish, crappie, and catfish, and they rely primarily on natural production. In the West, and that is where we live, trout is the dominant fresh water sport fish.

Fun Facts and a Diversion: How Does The "Put" in Put-And-Take Actually Work?

It must be springtime in the wilderness because it is raining fish from the sky. You could possibly be sitting by a sparkling lake high up in the California Sierras, contemplating solitude and beauty and communing with your inner-self, when a plane flies over and dumps thousands of trout into your previously mirror like lake, and that is the modern version of "put."

The Wild West of putting fish into waters where there were none began in earnest in the United States in the 1870s, when a fish train from the East, loaded with carp, could meet a fish train from the West, loaded with trout. Trout hatcheries multiplied. By the turn of the century, members of the Sierra Club packed trout into the high Sierras, and sportsmen's clubs and private citizens eagerly planted trout wherever a trout-free lake could be found. If a body of water did not dry out over the winter, it was fair game.

This free-for-all of planting trout occasionally took some strange turns. In 1964, Chuck Yaeger, then a colonel and famous test pilot for the US Air force, took one of his commanding officers, General Irving Branch, fishing at a lake full of golden trout that Yaeger had found during a serendipitous overflight 13,000 feet up in the Sierra Nevada. The general was so taken by the trout, "the color of old gold," and ordered Yaeger to introduce the trout into New Mexico. And so Operation Golden Trout was born. Yaeger and Branch commandeered a few US Air Force aircraft, enlisted some employees of the New Mexico Fish and Game department, and in the dark of night, as this was clearly illegal, absconded goldens from their Sierra lake, placed them into special oxygenated containers and onto a New Mexico State fisheries vehicle, packed vehicle and trout into a C130 cargo plane, and flew them to New Mexico where the trout were kept over the winter and eventually released into several lakes in an area where Branch planned to retire.

Originally, the goal of stocking fish was two-pronged. Fish should be introduced to areas where they could establish a wild population, and fish should be stocked annually to keep anglers happy. Over time, political pressure to keep anglers happy overrode all other potential outcomes.

No thought was given to any management, record keeping, follow-up, or considerations for the biota that existed already. Nobody really knew where, how, and when trout were being introduced. Biodiversity and ecosystems were not yet a gleam in managers' eyes. Once planes and helicopters started dropping trout, no lake was safe, and, occasionally, the wrong lake was stocked, but as long as the anglers were happy, all was as it should be.

The results were mixed. Trout are rapacious predators, and they hoovered up the existing invertebrates and amphibians and initially grew fast and big, a boon for the anglers. With time though, in many places, the trout ran out of food and had to stop growing.

Quite a few years ago, we visited Don Hall, then a graduate student in the high Sierras, whose goal it was to empty a lake of introduced brook trout and see what would happen. We fished the lake with a gillnet from an inflatable raft and lined up over 800 peculiarly shaped trout on the rocks. The fish were 10 years old and had outsized heads and a tiny body. They were starving. But there was another lake a little higher up where the brook trout did really well and grew to respectable size. Why? The answer to that will take a few more dedicated graduate students spending their summers doing experiments at 11,000 feet.

Beginning in the 1960s, concerns grew and are still ongoing with the destruction of biota, that is, any indigenous life form, and other deleterious effects of willy-nilly introductions of trout. There is some pushback now to stop stocking lakes where the overall biota might be damaged. Where no harm is done, keep on stocking. However, stocking is so entrenched in the public's mind, that fishing success is seen to depend on regular stocking, even though that may not be true at all.

But the trout in all its permutations, rainbow, golden, brown, cut throat, brook, etc. have been successfully domesticated and have conquered the freshwaters of the world (see Figure 7.1). They may outlive us all.

Their original range was the Western United States, Canada, and Eastern Russia.

Recreational fisheries, particularly introduction of exotic species such as trout to much of the world, has generated great fishing opportunities but has also done great damage to the native biota where the introductions have taken place. Modern biosecurity regulations in many countries would have prohibited almost all of the introductions that were done, and the free-for-all transplanting of species has largely come to an end.

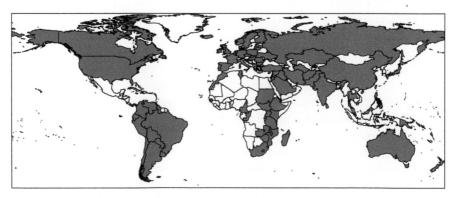

Figure 7.1 A map showing where rainbow trout have been introduced or are found naturally.

The end of the trout diversion and back to how to manage all kinds of recreational fisheries.

How Recreational Fisheries Are Managed

Who gets to fish is another management tool that comes in two contrasting versions in the world of recreational fisheries. In Europe and China, those who are willing to pay get to fish for freshwater fish that are generally privately owned. In almost all marine fisheries, anyone who wants to go fishing can go. Sometimes a license must be bought, and fees may be lower for locals than for outsiders, but on the whole, marine fisheries are essentially all open-access with no attempt to limit the number of anglers.

Few recreational fisheries are regulated by a total allowable catch—those I know of are almost exclusively in the United States under federal management using commercial management methods.

A big difference in the United States is that sport fish license fees and taxes on fishing equipment pay much of the cost of State sport fish management.

Let us explore three variations on the theme: (1) Recreational fisheries management of inland fisheries in Germany, (2) marine fishery for crab in Washington State (USA), and (3) inland fishing in China.

Germany

Commercial and recreational freshwater fisheries are usually governed using fishing rights systems, which are private-property systems that entail the right to catch wild fish, while being legally coupled to the duty to maintain stocks in near-natural conditions. Basically, there is a fishing right on any inland water, usually tied to ownership of the water body. Private owners of fishing rights entail individual commercial enterprises, angling clubs, angling associations, and private individuals. On the whole, larger bodies of water (rivers and large lakes) are publicly owned, whereas smaller lakes, artificial lakes (e.g., gravel-pit lakes) and ponds tend to be often private. Whether public or private, fishing rights exists on all waters, and, even if the water body is publicly owned, the fishing rights for that body of water are usually leased for 12-year periods to fisheries enterprises or associations or clubs of anglers. Whoever holds these rights can regulate who fishes, how big the take is, and what to stock, as long as these actions comply with minimum-standards set in state-specific fisheries legislation (e.g., the release of nonnative fishes is legally banned or state-wide minimum-length limits are set in the laws). The dominant rights holders are angling clubs or associations, and here the divided Germany is still evident. In the old West, it is dominantly small fishing clubs who hold the rights, whereas in the East, the rights are held by state-level organized fishing associations (where organized angling clubs themselves are not fishing rights holders). Anglers need to become members of either clubs or associations to be able to fish or they can buy tickets from commercial enterprises, and the revenue money is invested by the fishing rights holder into management and enforcement. The state government expects the rights holders to be able to manage their fisheries by means of some sort of formal training (e.g., for fisheries managers within organizations), and requires individual anglers to pass an examination that demonstrates their knowledge of fish biology, regulations, and appropriate fish handling.

Washington State

Dungeness crab is the absolutely best crab known to man (most certainly to the man writing this and to his wife who rewrites everything) and a highly valuable commercial and recreational fishery along the West Coast of the United States. Recreationally, it is managed as an open-access public resource by the Washington Department of Fish and Wildlife (WDFW). For the finer management details, we will use Penn Cove on Whidbey Island in Puget Sound, where our son occasionally invites us to go crabbing with him. Each individual license allows for two baited pots per day with a daily catch limit of five crab per license. Each licensee must be present on the boat. Only male crabs that have reached a specific carapace size can be kept; females must be returned to the water. The pots must be designed so that small crab can escape. The season is limited to July and August, and is closed on specific days of the week to give the crab a breather. WDFW monitors the catch based on the catch records that must be returned at the end of the season, and will reduce the season length and bag limits when abundance appears to be low. Regulations in the whole of Puget Sound can be tweaked according to the reporting of catch.

China

Freshwater recreational fishing in China is primarily a private business where people pay to fish in a pond, lake, or river. The body of water may be owned by local collectives or private individuals that are mostly small-scale cooperatives or family farms. Since there are few wild, self-sustaining fish available, most fish caught recreationally have come from hatcheries. The only real form of management is the stocking.

I have argued that sustainable fisheries management requires pristine habitats, the ability to monitor changes in abundance of the targeted fish population, and the ability to reduce fishing pressure when the population declines.

Monitoring the abundance of recreational fish in the ocean poses the same problems as it does for commercial fish, except that the targeted species are more commonly in coastal waters, and scientific surveys are rarer. Where there is a mixed commercial and recreational fishery, monitoring catch per day or catch per hour fished is mostly used for recreational fisheries, but can equally be used for the commercial catch.

The most challenging part is to estimate the catch and fishing effort. Fishing happens in many disparate places by many people out on the ocean or in inland lakes and streams. Information on how often someone went fishing and what and how much was caught is gathered from returned licenses, telephone interviews, mailed questionnaires, randomized estimates from counts at boat launches or aerial surveys, and spot checks at landing sites. Clearly, the estimate of catch and effort for recreational fisheries is imprecise compared with the numbers that come from commercial boats.

What to do when the warning signals go up such that a target fish population is in trouble in a fishery that values maximum long-term opportunity to fish and maximum effort? The best management tools here are to reduce catch without reducing effort by increasing the size limit, decreasing the bag limit, and, if needed, shortening the season and closing certain areas. Ideally, people can still go fishing but kill fewer fish.

Conflict between Recreational and Commercial Fishermen

When both commercial and recreational fishermen compete for the same resources, unreasonable amounts of passion and money are expended on lobbying in legislators' chambers and on protracted battles in court over who gets the fish.

Sport fishermen often have a real problem equating the few fish they take with the much larger catch of a commercial boat when it comes to the impact on the fish stock. And yet, one fish each caught by a thousand fishermen is the same number of dead fish as a thousand fish caught by one fishing boat, and, in many cases, the death toll from anglers is the same as that from commercial boats. But a thousand fishermen may have more political clout than ten fishing companies do, and consequently they sometimes succeed in eliminating or highly restricting the commercial fishery.

In 1994, Florida voters approved an amendment to the state constitution banning the commercial use of nets—effectively eliminating most commercial fisheries. This was a great victory for the sport fish industry, and, while it definitely had some conservation benefits, it did give the commercial fish to the recreational anglers. A similar ballot initiative was proposed in Washington State, but it was rejected by voters after many conservation groups pointed out that reallocating the commercial fish to recreational fishermen had nothing whatever to do with conservation.

Not content with reallocating catch, anglers are pushing to have certain species declared as game fish. Once the legislature has done that, all commercial harvesting of that newly minted game fish is prohibited. For years now, recreational fish lobbyists have been exerting effort and pressure to have the striped bass of the US East Coast declared a game fish, which would wrest it out of the clutches of commercial fishermen.

Meet the New Zealand Snapper—A Very Fine and Desired Fish

In northern New Zealand, the holy grail of sport fishing is known as the snapper. It is not technically a snapper but a sea bream, but notwithstanding, it is a lovely fish of a pleasing shape, a baby pink decorated with the odd blue dots, very tasty, and, because everybody wants to catch it, it causes the most ungodly management muddle.

The largest individual stock is located around Auckland, in the Bay of Plenty and east of North Cape, where the majority of New Zealanders live. The commercial total allowable commercial catch is 4,500 tons, and the most recent recreational catch is estimated to be about 3,800 tons, with 2,300 tons taken in the Hauraki Gulf, next to Auckland. The commercial fleet operates under an individual transferable quota (ITQ) system that caps the annual catch. Sport fishing regulations are deceptively simple and do not have a cap: a 30-cm minimum-size limit (MSL) and a seven fish-per-day-per-fisher daily limit, regulations that were implemented in April 2014. New Zealand does not require a license to fish recreationally in marine waters, and there are no reporting requirements.

The core of the snapper problem is the struggle between the commercial and the recreational fisheries. While the commercial fishery brings direct revenue into the country, the recreational fishery deems its fishery to be much more valuable to the country and, therefore, would like to see it increased. In the 1990s, the court ruled against any agreed share of the catch for New Zealand commercial fisheries. In other words, more recreational catch necessarily means less commercial catch. Given that the commercial catch is directly limited by ITQs, it is easy to set new rules for a smaller catch. The recreational catch, on the other hand, is loosely regulated through minimum legal size and daily bag limits. As can be expected, this approach is not overly successful. The most recent estimate for the actual recreational catch was 3,800 tons, whereas the stated "allowance" was only 3,050 tons. Consequently, it is the commercial fishermen who bear the brunt of catch reductions to conserve and increase the snapper population.

Putting the burden of conservation on the commercial fishery, we can expect the snapper population to increase, and that will mean a bigger share for the recreational fishery unless the rules change. Worse yet, if more New Zealanders decide to fish and fish more often, the recreational catch share will increase again to the detriment of commercial fishermen whose share will dwindle. One might be forgiven for thinking that the drivers of the conservation efforts ought to benefit.

The sport fishermen see it differently. Snapper is a valuable resource enjoyed by hundreds of thousands. In their opinion, the ITQs benefit the few companies who own them and a few hundred commercial fishermen who catch and process the fish.

If those who set up the New Zealand ITQ system had had the wisdom to include recreational fishing allocations at the beginning, the government could buy ITQs from commercial fishermen and in turn, increase the recreational share. Money would solve the allocation dispute and put to rest the acrimony and political bickering. And, there could even be a license fee for marine recreational fishermen, providing the money needed to buy that quota. Unfortunately, the possibility of this eminently sensible solution is politically most unpopular. There also exists an important recreational charter boat fleet, where the operators guarantee a snapper for each client, and who constitute as much of a business operation as does the commercial fishing fleet. However, it is deemed by the government to be sport fishing, and, therefore, the participants need no license to fish. Since 2010, recreational charter operators have been required to register with the government and report their catch from every paying trip, so there is a distinction between these two components of the snapper recreational fishery. The existence of the charter boat fishery definitely also blurs the line between recreational and commercial fisheries.

●●●

FURTHER READING

An overview of recreational fishing participation. Arlinghaus R, Tillner R, and Bork M. 2015. Explaining participation rates in recreational fishing across industrialised countries. Fisheries Management and Ecology 22: 45–55.

Another overview of recreational fisheries. Cooke S J and Cowx I G. 2004. The role of recreational fishing in global fish crises. Bioscience 54: 857–9.

A comparison of management in the United States and Germany. Daedlow K, Beard T D Jr., and Arlinghaus R. 2011. A property rights-based view on management of inland recreational fisheries: Contrasting common and public fishing rights regimes in Germany and the United States. In: American Fisheries Society Symposium Bethesda, MD: American Fisheries Society, 13–38.

FAO. Technical guidelines for responsible fisheries: Recreational fisheries. No. 13.

Some estimates of the extent of recreational fisheries. World Bank. 2012. Hidden harvest: The global contribution of capture fisheries. Washington, DC: World Bank.

Chinese recreational fisheries. Shen J. 2008. Current status and challenges facing recreational fishing in the People's Republic of China. Global challenges in recreational fisheries. Oxford: Blackwell Science, 18–21.

CHAPTER 8

Freshwater Fisheries

W orld fisheries = marine fisheries, right? Be it conservation concerns associated with overfishing, food production of large fisheries, or the economic potential of a nation's fisheries, marine fisheries dominate the political and scientific discussion. Yet it is possible that freshwater fisheries are more important for food security. They involve more people and may even be more valuable than marine fisheries. The landing statistics of the UN Food and Agriculture Organization (FAO) show about 15 million tons for freshwater fisheries. Yet the underreporting of catch is so vast and given that most are small, poorly studied, many not documented at all, this makes for a lot of "dark" numbers. Freshwater landings may well exceed the 64 million tons reported for marine fisheries. In other words, while freshwater fisheries are too big to ignore, they are certainly neglected.

We are not just talking of recreational anglers here, but of a widely scattered and immensely diverse range of activities, starting at the industrial end with the export fisheries for Nile perch on Lake Victoria, to the classic small-scale subsistence and artisanal fisheries in Asia and Africa and finally the recreational fisheries of the developed world. Their targets are both species that spend their entire lives in freshwater and those that migrate between salt water and freshwater.

It is important to note that freshwater fisheries have very high participation by women, in terms of subsistence fishing, but also on the processing and marketing side of commercial fisheries.

Yield and Regulations

Perhaps the most interesting statistic about freshwater fisheries is their potential yield. Based on surface area and known productivity, there are estimates that they could produce 72 million tons of harvest,[1] that is more than present marine fisheries bring. Given how truly sketchy the catch statistics are, it is entirely possible that the real catch is already that high.

Ocean Recovery: a sustainable future for global fisheries? Ray Hilborn and Ulrike Hilborn, Oxford University Press (2019). © Ray Hilborn and Ulrike Hilborn 2019. DOI: 10.1093/oso/9780198839767.001.0001

Fishing in freshwater goes back to the dawn of time and has been haphazardly regulated every so often. There were fishing regulations in China as early as the eleventh century BC. An Etruscan bas-relief of a sturgeon shows the minimum size for sale. Fisheries regulations were included in early versions of the Magna Carta, and in the 1600s, the French regulated minimum sizes of fish caught in the Loire and Seine rivers. Not much has changed, haphazardly regulated still applies today. There have actually been some scientific investigations and even a few attempts at regulation. However, most freshwater fisheries, except recreational, are now in Asia and Africa, where there simply is no overriding management framework, and harvest and stocking are not at all regulated by any central authorities, with the exception of the odd attempt by community or regional systems. There used to be large freshwater commercial fisheries in the United States and Europe, particularly in the Laurentian Great Lakes, but these are now primarily recreational fisheries.

Table 8.1 lists many of the largest river basins in the tropical world where freshwater fish production is very important now or there is great potential for the future. This list includes river basins with over 2 billion people.

Dams, Pollution, Water Withdrawals, and Exotic Species

Freshwater fisheries have their own problems, but managing harvest is only one of them; there are also dams, irrigation, and exotic species.

Spread all around the world are over 57,000 large dams that are more than 15 meters high. They produce hydroelectric power and flood control, with the added

Table 8.1 *Largest River Basins in the Tropical World*

River Basin	Area (million km²)	Population millions
Amazon	6.1	28
Congo	3.7	120
Nile	3.3	257
Rio de la Plata	3.2	100
Niger	2.3	100
Yangtze	1.8	430
Ganges- Brahmaputra	1.7	630
Zambezi	1.3	32
Indus	1.1	300
Mekong	0.8	60
Yellow River	0.8	189

benefit of river transport and irrigation in some cases. But dams also block the passage of fish, flood habitat, alter flow and temperature regimes, and facilitate water withdrawals for agriculture. Because of the need for electricity, an additional 3,700 large dam projects were planned in 2014.

Agricultural runoff and human and industrial waste are the main causes of pollution of rivers and lakes. With pollution and dam construction for mills, the industrial revolution brought the first fish free streams to the English countryside. It was not until the twentieth century that modern pollution control cleaned up the Thames and other rivers and, sure enough, the native fishes came back but in much smaller numbers.

Irrigation is both a blessing and a curse. Agriculture is thirsty and needs freshwater drawn from rivers and lakes. The fish are left with a diminished supply of clean water and a backwash full of fertilizer and pesticides.

The Aral Sea: A Harsh Lesson

"The people destroyed the sea and then nature took revenge on the people."[2]

The most devastating twentieth century example of irrigation gone rampant is the Aral Sea in Central Asia, once the fourth-largest lake on earth. Surrounded by desert, it is fed by two rivers, the Syr Darya from the north and the Amu Darya from the south, but it has no outflow. It was a highly productive inland sea with a thriving fisheries community that produced 48,000 tons of fish in 1957. In 1987, there was no more commercial catch and the port city of Aralsk found itself 12 miles from the lakeshore.

In 1960, the Soviets decided to make the desert bloom with cotton. The irrigation projects took almost all of the water of both rivers, and soon the lake began to shrink and kept shrinking. The fishermen had to haul their boats farther and farther to the receding water that became more and more salty. Only flounder could tolerate high salinity; the rest of the fish disappeared. The lake had been cut in two: a dry south with a trickle and a small lake in the north. Dust from the dry lakebed, polluted with chemical fertilizers and pesticides, became a human health hazard and settled on the fields, requiring ever more river water to flush out the salt. Without the mitigating large body of water, the climate changed: summers got hotter and winters got colder.

Once the Soviets were gone, the countries of Kazakhstan and Uzbekistan were left with the devastation of the Aral Sea and a poverty-stricken population. Kazakhstan, with the help of the World Bank, repaired the dikes and built an 8-mile dam across the old lakebed to let the Syr Darya fill up the remaining northern lake. This intervention was unexpectedly successful. The fish came back and the legal fishing limit for 2018 was set at 8,200 tons, which does not account for a considerable illegal catch.

Uzbekistan's share of the Aral Sea is a dry, dusty, poisoned lakebed.

Introducing Exotic Species: Something We Just Cannot Resist

The biggest transformation of freshwater fish communities happens with the introduction of exotic species. Tilapia and Nile perch devastated the highly diverse native cichlid community in Lake Victoria. Of the lake's 500 species of cichlids, 200 are estimated to have totally disappeared by the late 1980s.[3] Cichlids are a vital source of protein for the people who live around Lake Victoria, but they had almost disappeared from the catch in the small-scale local fisheries. Recently, Nile perch have been fished down, and some cichlids are coming back, but the latest worry is pollution from the ever-increasing population around the lake.

Where Do Most of the Freshwater Fish Come From?

The bulk of the catch of freshwater fish comes from the large river systems in tropical Asia and Africa, and there is potential from the big rivers in South America. Let us look at some of the truly big tropical freshwater systems.

The Mekong

The Mekong is the twelfth longest river in the world. It flows from the Tibetan Plateau through China, Myanmar, Laos, Thailand, Cambodia, and Vietnam, a watershed that supports around 60 million people (Figure 8.1).

The most recent estimates of the basin's annual fish production are 2,642,000 metric tons by capture fisheries plus another 260,000 MT by aquaculture. Subtracting for export, the annual per capita fish consumption in the Mekong basin is estimated to be somewhere between 25 to 50 kilograms per person, the highest in the world. Obviously, the fish of the Mekong are an absolutely vital part of people's nutrition.

Few make a full time living fishing on the Mekong though. It is people who call themselves poor or very poor who survive through a mix of part-time fishing, some agriculture, logging, and perhaps a bit of aquaculture. According to surveys, people see two major threats, overfishing, and loss of habitat attributable to pollution and agricultural development.

However, there are already more than fifty hydroelectric dams, and more than one hundred new ones are planned. Several of the very large dams planned or under construction will block the migration of some of the most important species of fish, which has set off alarms among environmental groups. If all of the big dams planned for the lower Mekong are actually built, fish catches have been estimated to go down by 42 percent.

If there is any form of fisheries management, it is strictly local when communities organize (see chapter 3), but on the whole, fisheries and the integrated aquaculture are largely unmanaged and poorly documented. Even local efforts such as in the Tonle Sap region of the river in Cambodia are not entirely successful. About 6 years ago, there was a huge policy shift to a community-based management system. The result has been a massive increase in effort and reports of declining catches.

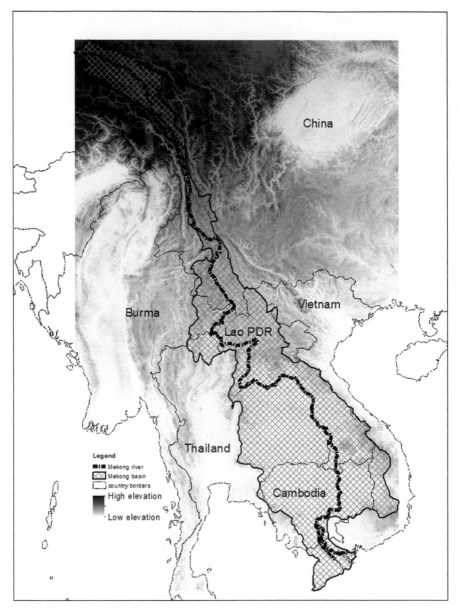

Figure 8.1 Map of the Mekong River basin.

The Amazon

The Amazon is the largest river in the world (Figure 8.2), and yet it produces a relatively modest amount of catch each year because of very low population density.[4] In the upper reaches, fish are the dominant source of animal protein, but, currently, most of the fisheries are on the lower river, where both artisanal and commercial

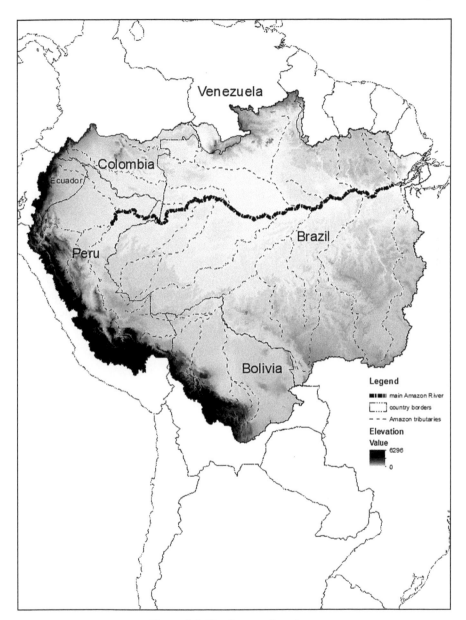

Figure 8.2 The Amazon River basin.

fishing coexist. For those, some co-management systems do seem to improve fisheries performance.

Because of its size, dams on the Amazon River itself are not in contemplation, but there is a dam construction boom on the tributaries. The Belo Monte Dam on the Xingu River, at capacity in 2015, is the fourth-largest hydroelectric project in the world, and many similar projects are slated for other tributaries. Displacement of indigenous peoples through these massive dam projects is cause for great concern internally and internationally.

Aside from Brazil, Peru and Bolivia also have plans for major dam projects on Amazonian tributaries. In total 428 individual dams are planned for the Amazon basin, and 140 are either complete or under construction.[5]

Lake Victoria

Lake Victoria, the second-largest freshwater lake in the world by area, has the single largest freshwater fishery. The arrival of the railway in 1908 opened up the lake to distant markets, and a commercial fishery developed rapidly. By 1927, overexploitation of the native fishes was already enough of a concern to begin scientific surveys that suggested a minimum mesh size and a lake-wide authority to collect statistics and control fishing effort (Figure 8.3).

Then came Tilapia and Nile perch. High-value whitefish that are mostly exported, they were introduced in the 1950s and entirely transformed Lake Victoria fisheries. With jobs and money, the population around the lake increased, and so did the fishing effort, and down went the cichlids.

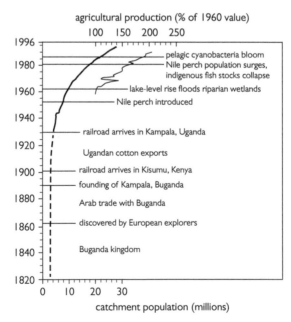

Figure 8.3 A timeline of development of Lake Victoria showing growth of human population, agricultural production, and significant events. Redrawn from Verschuren D, Johnson T C, Kling H J, Edgington D N, Leavitt P R, Brown E T, Talbot M R, Hecky R E. 2002. History and timing of human impact on Lake Victoria, East Africa. Proceedings of the Royal Society of London B: Biological Sciences 269: 289–94.

Kenya, Tanzania, and Uganda border the lake. All three had been British colonies that pretty much kept their British-type administration after independence, and they attempted to manage the Nile perch fishery along European lines with stock assessment and limiting effort. Regulations include limiting the size of nets and their mesh, licenses, and enforcement. But despite regulations, catch data revealed that as much as 99 percent of females were below the size at maturity. Attempts to lower effort have failed consistently, and abundance continues to go down. The latest measures to halt this decline are enforcement of vessel registration and limiting entry of more boats. The cichlids, on the other hand, profited from all that overfishing and repopulated some of their old habitat.

Nile perch is much too expensive for most people who live around the lake. Artisanal and subsistence fishers go out in small craft for small pelagic fishes known as dagaa, with small-scale gear that now includes mosquito nets. The fish are largely dried or smoked and are the cheapest form of animal protein in the region. Dagaa fishing is ubiquitous and effectively unregulated.

With the rapidly growing population, more land has been converted to agriculture. More people and more agriculture mean more deforestation, sedimentation, pollution, and runoff into the lake. Those added nutrients caused the phytoplankton to explode and create anoxic dead zones on the bottom of the lake. Bluegreen algae bloomed and made such thick mats on the lake that no fish could live underneath them.

Moving Forward on Inland Fisheries Management

We cannot overstress how important inland fisheries are to food security, as a source of micronutrients and protein for decent nutrition, and a means to lift people from poverty or at least a fallback in times of hardship. Nor can we overstress the appalling lack of knowledge, scientific investigations, and management. In recognition thereof, FAO[6] laid out ten steps to alleviate this dire situation.

- Step 1: Improve the assessment of biological production to enable science-based management.
- Step 2: Correctly value inland aquatic ecosystems.
- Step 3: Promote the nutritional value of inland fisheries.
- Step 4: Develop and improve science-based approaches to fishery management.
- Step 5: Improve communication among freshwater users.
- Step 6: Improve governance, especially for shared waterbodies.
- Step 7: Develop collaborative approaches to cross-sectoral integration in development agendas.
- Step 8: Respect equity and rights of stakeholders.
- Step 9: Make aquaculture an important ally.
- Step 10: Develop an action plan for global inland fisheries.

These ten steps illustrate how much in the dark we really are. We know nothing definite about the value of the fishery or even how many people participate, never mind the status of the stocks. There is also much uncertainty as to the need for collaborative management. What is certain, though, is that none of the countries where inland fisheries are important has the potential for national-scale, top-down management. Therefore, if there is to be any effective management of these vital resources, it must come from community-based co-management.

In summary, even though inland fisheries are vastly important, they are by far the most poorly understood and barely managed.

FURTHER READING

An overview of freshwater fisheries. Welcomme R. 2016. Inland fisheries: Past, present, and future. In: Freshwater, fish and the future; proceedings of the global cross-sectoral conference. In: Taylor W W, Bartley D M, Goddard C I, et al. (eds). Rome: Food and Agriculture Organization of the United Nations.

The estimate of potential freshwater fisheries production. Lymer D, Marttin F, Marmulla G, et al. 2016. A global estimate of theoretical annual inland capture fisheries harvest. In: Taylor W W, Bartley D M, Goddard C I, et al. (eds). Freshwater, fish and the future; proceedings of the global cross-sectoral conference. Rome: Food and Agriculture Organization of the United Nations.

The steps needed to advance freshwater fisheries management. Taylor W W and Bartley D M. 2016. Call to action–the "Rome Declaration": Ten steps to responsible inland fisheries. Fisheries 41: 269–9.

The story of the Aral Sea. Dene-Hern Chen. 2018. Once written off for dead, the Aral Sea is now full of life. National Geographic (March): https://news.nationalgeographic.com /2018/03/north-aral-sea-restoration-fish-kazakhstan/.

CHAPTER 9

Mixed-Species Fishing and Bycatch

Overview of Mixed-Species Fisheries

Fish rarely arrange themselves in one-species aggregates. When a net or a longline comes up, not everything that is caught is desired. It will usually be a mixed bag of the target species: the ones you really want and other species that are not always of any commercial value or fish that are too small or species that are not fish at all. These nontarget species can turn up as birds, dolphins, turtles, seals, or sharks. In other words, a haul of mixed species brings a whole lot of problems with it.

People have fished ever since prehistoric bands settled on a lake or a river or on an ocean shore, and, from very early on, they used three basic types of gear: a hook on a line, a net, or a trap. And even though we have vastly refined our fishing gear, this triad of design is what we still use now. Hook-and-line gear can be a single baited hook dropped from a boat with a weight on the end or it can be an industrial longline, maybe 20 miles long with hundreds of baited hooks. Native Americans use small hand held dip-nets to catch salmon when they swim upstream. Tuna boats use gigantic purse seines over a mile long and hundreds of feet deep to encircle schools of tuna. Trawl nets catch a wide range of species and account for one-quarter of all fish landings. Fish traps can be simple woven baskets used in many traditional fisheries, or the 800-pound monsters for King crab seen on "Deadliest Catch."

There are some fishing methods that are far more selective than the average mixed-species fishery by the kind of gear they use and when and where they fish. The net fisheries for salmon in Alaska rarely catch anything but salmon, and lobster pots in New Zealand catch mostly lobster, but these are the exceptions rather than the rule.

When multiple species are caught, problems arise because not all species should be harvested at the same rate, and we do not want to catch some species at all.

Scientists try to estimate the rates of harvest for an individual species that would produce maximum sustainable yield. The fraction of the population that

Ocean Recovery: a sustainable future for global fisheries? Ray Hilborn and Ulrike Hilborn, Oxford University Press (2019). © Ray Hilborn and Ulrike Hilborn 2019. DOI: 10.1093/oso/9780198839767.001.0001

can safely be harvested depends primarily on the life span of the fish. Species that live a long time do not reproduce that fast and need to be harvested at a very low rate, whereas short-lived fish, usually small and plentiful, can be harvested at a much higher rate. For example, a 3-year old Peruvian anchoveta is a venerable grandmother and very rare. This is also the largest fishery in the world that has given us catches of up to 10 million tons, and that is more than 10 percent of the global fish catch. Because anchoveta live fast, reproduce quickly, and die young, we can fish almost half of the existing stock and still end up with the desired maximum sustainable yield.

In, contrast, the slow and steady orange roughy, found in deep waters around much of the temperate ocean, do not contemplate connubial bliss until they are 20 to 30 years old and can live to be 100. For orange roughy, a 3 to 5 percent harvest rate per year is the limit.

What happens when short- and long-lived species swim together in the same area? Not that orange roughy would ever cross paths with anchoveta, but other species certainly do use the same real estate. Then it gets tricky. When you fish the short-lived species hard enough to get the best possible yield, you overharvest the long-lived ones.

Let Us Explore A Few Examples

First, we will go to Kenya's artisanal fisheries. You have already met Tim McClanahan who is working with the small-scale fishermen to try to improve their livelihoods by better fisheries management.

The coastal Kenyans use every gear available, beach seines (large nets that surround fish), hook and line, other kinds of nets, spear guns and traps, with which they catch fish from 214 species of twenty-three families of fish. The lifespan of the most commonly caught fish ranges from 2 to 3 years up to 20 or more. The long-lived fish are larger and more valuable, but have disappeared or are extremely rare. All fish are heavily exploited. Tim works with the fishermen to rebuild the abundance of the long-lived fish to create more value by eliminating the use of beach seines that are the most indiscriminate in what they catch.

214 different species may seem like a lot but it is typical of fisheries in the tropics where species diversity is very high.

Let us take another view of China's industrial fisheries that also catch a wide range of species of different productivity mostly with bottom trawls. Nets are dragged along the bottom and almost every living thing is caught. Fishing pressure in China is truly heavy: there are estimates that in coastal waters a trawl net may pass eighteen times a year over the same spot. As in coastal Kenya, long-lived species are largely gone, and the catch is dominated by short-lived species that can sustain such high exploitation. A fascinating consequence of such super pressure on fish is the change in life history of some of the most important species. 14 percent of Chinese landings are largehead hairtail that used to be a large species that matured at 2 or 3 years. What with all that fishing pressure for the last few decades, the largehead hairtail has become an annual species, and it now matures in its first year of life, and only few live to be 2 or 3 years old. Evolution before our very eyes.

The economics of mixed-stock fisheries are also a challenge. If we fish at a low enough rate to allow long-lived species to survive and be sustained, not all of the potential short-lived species can be caught and we will forgo part of the potential catch. In general, short-lived fish species are more abundant than long-lived species, but long-lived species are more valuable. Is there a middle ground where we can have both?

It all depends on your ultimate goals. If food security is our primary goal, we would choose the highest possible sustainable yield in tonnage from the ecosystem. If protecting livelihoods is our primary goal, we would go for the money to get the best sustainable value of the catch. If our goal is to assure that all species are maintained at high abundance, we would decide on conservation measures, sometimes called weak-stock management, to adjust fishing to assure the abundance of the stock with the lowest productivity—usually the longest lived species.

The mixed bottom-trawl fishery on the West Coast of the United States is a well-studied example of weak-stock management.[1] Up to one hundred different species have been caught, and thirty-four of them have been the subject of detailed scientific analysis. Theoretically, the yield of the seven long-lived species could be maximized at annual harvest rates of less than 3 percent, of another fourteen species at 3 to 10 percent, and of thirteen species at 10 percent. The total estimated sustainable yield of this mix of stocks, provided that each could be harvested optimally, is a little over 60,000 tons. The seven "weak stocks" with low desired harvest rates constitute only 14 percent of the total possible yield, whereas the strongest stocks at harvest rates over 10 percent could produce 42 percent of the potential yield.

In order to maximize the total yield from this mix of stocks, a harvest rate of about 12 percent would be appropriate—a compromise between underharvesting the strong stocks and overharvesting the weak stocks. However, if we wanted to assure that each stock is maintained at maximum yield or higher, the harvest rate goes down to 1 percent with a much lower yield. Because US fisheries law has mandated weak-stock management since 1996, harvest rates have been about 1 percent per year, and the catch has declined from about 60,000 metric tons per year to about 20,000.

When you throw natural variability in ecosystems into the mix, stocks that normally would be subjected to an average sustainable harvest rate may naturally fluctuate in abundance at different times and become weak stocks because of temporary low abundance rather than a low sustainable harvest rate.

Mixed-species management can also tinker with gear to achieve the desired outcome of avoiding weak stocks and catching strong ones. Gear design and how, when, and where gear is used can make a difference.

A Closer Look

Figure 9.1 shows the total yield across all the groundfish stocks on the US West Coast against what fraction are harvested as a thick black line. Total sustainable yield would be maximized by harvesting 12 percent of the abundance.

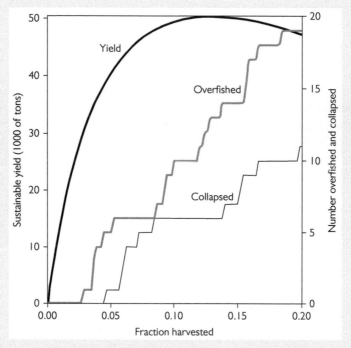

Figure 9.1 The relationship between the fraction harvested and the potential yield (black line). The solid gray line shows the number of stocks that would be considered overfished.

To have no overfished stocks the fraction harvested would need to be about 3 percent. To have no stocks collapsed the fraction harvested would have to be less than 5 percent (thin solid line).

This graph shows the inevitable trade-off between maintaining all stocks at high abundance (none overfished) and the actual yield from a mixed-stock fishery.

Fishermen know where and when species are usually caught and can avoid catching weak stocks by adjusting time and place. They can also differentially harvest individual stocks out of the mix because of where they choose to fish.

Rockfish prefer to live on the rocky bottom of the coast of California and are the weak stocks of the US West Coast fisheries. Managers have closed these areas, called "rockfish conservation areas," to bottom trawling altogether.

Discarding

Discarding happens when nontarget species of no commercial value are caught and dumped overboard.

The most recent estimate is that about 8 percent of fish caught are discarded. In the shrimp fisheries, commonly anything that is not shrimp is discarded, and that anything is 80 percent of what is caught.

In recent decades, market forces and technology have put a stop to traditional discarding by the mixed-stock fisheries of South East Asia. Now essentially all of the catch is kept. The best and most valued species are sold to local or external markets. Less desirable species are turned into surimi or fish paste, and the dregs, often in the form of fish slurry at the bottom of the hold, become aquaculture feed.

There is also regulatory discarding, that is discarding by law. The point of it is to discourage fishermen to target certain species they might catch occasionally. That species must be thrown away. Or when the catch limit for a certain species is reached, whatever has been caught beyond that limit must also be thrown out.

This kind of regulatory discarding got a high-profile airing when UK celebrity chef Hugh Fearnley-Whittingstall talked about the discarding of cod because the cod limit had been reached, but the fishermen still had quota for haddock and other species and wanted to keep fishing. Throwing out potential food did not look good on TV. The European Union Commission responded with banning all discards, which meant that once the quota for cod was reached, the boat must stop fishing. As of 2018, cod is the weak-stock of the North Sea and has become the choke species. The EU Commission has effectively changed to weak-stock management, and, if the discard ban can be enforced, a significant decline in total catch from its waters is inevitable.

Marine Mammals, Birds, Turtles, and Sharks—Very Problematic Bycatch

The biggest bycatch challenge for many fisheries and the biggest conservation concern is the accidental catching of marine mammals, birds, turtles, and sharks. On the whole, whenever such a bycatch issue was identified, and regulations put in place and enforced to reduce the bycatch, design changes in gear and how it is operated have reduced the rate of bycatch dramatically.

Sadly, for a dozen marine mammal species, this has not been the case, and these are declining rapidly. The most notable is the Mexican porpoise, the vaquita, now on the brink of extinction because of bycatch in legal shrimp and recently in illegal totoaba gill net fisheries.

Yellowfin tuna in the Eastern tropical Pacific were originally caught by pole and line boats. Once a school of tuna was found, small bait fish were thrown overboard to put the school into a feeding frenzy, and then unbaited hooks on the end of a short line attached to a pole would catch the frenzied tuna who would bite on anything to be hauled on board. It was hard work for large crews to catch heavy tuna one at a time. Then came the era of the "super seiner," a much larger version of the purse seiners used in many coastal fisheries to catch sardines and herring. 200 to 300 feet long, a super seiner could surround an entire school of tuna with a mile-long net, and the technology spread rapidly from its home port near Los Angeles. But dolphins swim with schools of tuna because they feed on the same small fish the tuna are chasing. Caught in the encircling nets, the dolphins drowned before they could be disentangled. There were probably hundreds of thousands of them caught, drowned, and crushed each year, and their population declined rapidly. In 1988, biologist Samuel LaBudde worked as a crewman on a Panamanian tuna seiner and with a video camera recorded the horrifying images of hundreds of dolphins dying in tuna nets. Public outcry was great, and soon public pressure ratcheted up to stop the killing.

In response, fishermen developed a procedure called "back down." Before the net is hauled on board, part of it is pushed below the surface, and a crew in speedboats herds the dolphins out of the net. Once clear of dolphins, the tuna are hauled onboard. The technique was remarkably successful. Dolphin mortality declined from hundreds of thousands to a few thousand, and the dolphin population began to recover. Back down takes a few hours more, but it is well worth it to the tuna fleet to almost eliminate killing dolphins.

A current issue of great concern is entanglement of whales in the lines that connect lobster and crab pots to the floats on the surface.

Marine turtles are killed by shrimp boats when they are caught in a bottom trawl and cannot come to the surface to breathe. Tens of thousands were thought to have been drowned before US government scientists developed turtle excluder devices (TEDs) that kick the turtle out of the net by a grid of bars, but let the shrimp through. Voluntary use of TEDs was slow in coming because the device also slightly reduces the shrimp catch, and, consequently, their use has been legally mandated.

Because they feed on the bait, turtles are also caught in longline fisheries. The traditional longline hook is called a "J hook." Bending the hooks into something like a circle proved a winner because turtles are less likely to bite, and even less likely to swallow such a hook. The switch to circle hooks by the Hawaii longline fleet reduced the catch of leatherback turtles by 84 percent and loggerhead turtles by 95 percent. In US Atlantic longline fisheries, the circle hooks were less effective, reducing catch by 64 percent for leatherback and 55 percent for loggerheads.

Seabirds will dive for the baited hooks when longlines are played out from the vessel. If they are caught on the hook, they will drown as the line sinks. There are now two clever solutions to the bird problem. One, the "tori line" is a line tied to a float and hung with long streamers to scare the birds away from the baited line while it is being let out. Tori lines have reduced seabird bycatch by 70 to 75 percent. Even more effective was a change in vessel design. Rather than letting the baited line out over the stern, the longline goes out from the bottom of the boat, deep enough that the birds cannot see the bait.

Fishermen are ingenious, and, if there are enough incentives, they can usually find a way to greatly reduce the bycatch of birds, mammals, and turtles. But so far, only a few countries have such incentives.

••

FURTHER READING

A big-picture look at bycatch and discarding. Hall M A, Alverson D L, and Metuzals K I. 2000. Bycatch: Problems and solutions. Marine Pollution Bulletin 41: 204–19.

Another big-picture look at discarding. Kelleher K. 2005. Discards in the world's marine fisheries: an update. Rome: Food & Agriculture Organization of the United Nations.

An overview of the sea turtle problem. National Research Council, 1990. Decline of the sea turtles: causes and prevention. National Academies Press.

The development of bycatch reduction methods for longline turtle interactions. Swimmer Y, Gutierrez A, Bigelow K, Barceló C, Schroeder B, Keene K, Shattenkirk K, et al. 2017. Sea turtle bycatch mitigation in US longline fisheries. Frontiers in Marine Science 4: 260.

A detailed look at the tuna dolphin issue. Hall M A. 1998. An ecological view of the tuna-dolphin problem: impacts and trade-offs. Reviews in Fish Biology and Fisheries 8: 1–34.

Detail on the coastal Kenyan mixed-stock fisheries. Hicks C C, and McClanahan T R. 2012. Assessing gear modifications needed to optimize yields in a heavily exploited, multi-species, seagrass and coral reef fishery. PLoS ONE 7(5): e36022.

Details on the US West Coast mixed-stock fishery. Hilborn R, Stewart I J, Branch T A, and Jensen O P. 2012. Defining trade-offs among conservation of species diversity abundances, profitability, and food security in the California Current bottom-trawl fishery. Conservation Biology 26: 257–66.

On coastal Kenyan fisheries. McClanahan T, and Humphries A. 2012. Differential and slow life-history responses of fishes to coral reef closures. Marine Ecology Progress Series 469: 121–31.

A sociologist's look at the San Diego tuna fishery. Orbach M K. 1977. Hunters, seamen, and entrepreneurs: The tuna seinermen of San Diego. Berkeley: University of California Press.

A detailed look at Chinese mixed-stock fisheries. Szuwalski C S, Burgess M G, Costello C, and Gaines S D. 2016. High fishery catches through trophic cascades in China. Proceedings of the National Academy of Sciences 114(4): 717–21.

CHAPTER 10

Bottom Trawling

The Perfect Villain?

Bottom trawling destroys the ocean! Rapists and pillagers of the sea! Trawlers flatten everything in their path! Clearcutting the ocean floor! Was there ever a more-deserving villain than trawling?

By the simple act of fishing the ocean floor, that ocean floor is changed. Whether it be plants or animals, the physical act of fishing alters both species composition and the ecosystem on ocean floors that are as varied in their appearance as the land. From mud to sand, to pebbles and rocks, to the glory of coral full of life, each one has its very own mix of inhabitants that is affected by fishing.

Food from the sea floor comes to us mostly through bottom trawling. It does exactly what it says—it drags a net across the bottom of the ocean and brings up roughly 25 percent of the global fish catch. Dragging anything along the seafloor will kill species that are not intended to be caught. Species that live on top of the sediment are much more vulnerable to fishing gear than are those buried in the sediment, but no matter, bottom trawling by its very nature catches most species in the path of the net, be they target or not.

Environmental NGOs love to hate bottom trawls. Many of their fundraising appeals are based on past practices, when the sea floor was made safer for fishing by dragging heavy chains between two boats to clear any rocks or obstructions that might snag the net and cause expensive damage. This was called "preparing the ground."

With Greenpeace in the vanguard, there has been a concerted push in Europe for banning bottom trawling, leading to the European Commission's ban on bottom trawling in EU waters deeper than 800 meters. As a result of this relentless vilification, many seafood rating systems will "red list" any fish caught by a bottom trawl. In 2016, however, the Seafood Watch program broke new ground by removing bottom trawl fish caught off the West Coast of the United States from their red list and labeled them a "good alternative" on their yellow list. The Marine Stewardship Council has certified bottom trawl caught fish of many fisheries including the South African hake fishery, the New Zealand hoki fishery, the

Ocean Recovery: a sustainable future for global fisheries? Ray Hilborn and Ulrike Hilborn, Oxford University Press (2019). © Ray Hilborn and Ulrike Hilborn 2019. DOI: 10.1093/oso/9780198839767.001.0001

Alaska and Russian pollock fisheries, and a range of rockfish and flatfish off the US West Coast.

With so much contention and whipsawing of public opinion whether bottom trawling is the devil or a partially fallen angel of industrial fishing, let us take a closer look at its extent around the world and its impact on the target species that live on the bottom of the sea. We considered nontarget species in Chapter 9 on mixed-species fisheries and bycatch.

But before we get into the details, it must be emphasized that trawling effort has gone down considerably over the last 20 years in much of the world. Developed countries have deliberately reduced trawling in their effort to halt overfishing and rebuild weak stocks.

Many of the results in this chapter come from a recent 6-year study called *Trawling Best Practices* or TBP, initiated by the Packard and Walton Foundations in the United States, with subsequent funding from CSIRO, the Australian Research agency, the Food and Agriculture Organization of the United Nations, a number of fishing companies, and the US government. I was one of the three co-leaders of this project.

What Does a Bottom Trawl Actually Look Like?

They come in many designs, but the most common is the *otter trawl*. A heavy metal wing-like structure known as "otter boards," or doors, spreads the net wide while its weight keeps the net on the bottom. A *pair trawl* uses two boats, one attached to each side of the mouth of the net to keep it open. A *beam trawl* uses a solid metal bar to hold the net open and on the bottom. A very different type of bottom gear is the scallop dredge that has heavy metal teeth to dig into the bottom and reach underneath the scallops that are buried in the mud or sand.

Trawl Footprints

In 1998, a paper appeared in the journal *Conservation Biology* called "Disturbance of the Seabed by Mobile Fishing Gear: A Comparison to Forest Clearcutting."[1] It stated, "The frequency of trawling (in percentage of the continental shelf trawled per year)…annually covering an area equivalent to perhaps half of the world's continental shelf, or 150 times the land area that is clear-cut yearly." In other words, we are led to believe that not only is the area trawled each year roughly equivalent to cutting down the entire Amazonian Rainforest, but that the physical action of trawls is similar to that of clearcutting—everything on the bottom is wiped out by the passage of a trawl. If this kind of clear-cut happened for a few years, it stands to reason that all the seafloor would soon turn into a lifeless desert, hence the authors strongly advocated for banning bottom trawling. Notwithstanding their argument, clearcutting obviously did not happen since bottom trawling still provides us with 20 percent of all seafood.

Mapping the Trawl Footprint: If It Is Not Clearcutting, What Is It?

It is really difficult to actually map the footprint of trawling. Luckily, great strides have been made in the last 2 decades. In 2002, the US National Research Council reviewed the impacts of bottom trawling in the United States,[2] using trawl effort reported in statistical areas that ranged in size from a low of 7.3 square nautical miles (about 25 square kilometers) off Alaska to a high of 705 square nautical miles in the Mid-Atlantic region. Since then, much more detailed data have become available that may come from start- and end-locations from GPS or from satellite monitoring systems, called VMS, that automatically report the location of a vessel.

These new, high-resolution data give us a much better idea of the actual trawl footprint. To understand why let us look at a particular example.

Trevor Branch developed mathematical models of the orange roughy fishery in New Zealand for his Master's degree and moved on to the behavior of the trawl fishing fleet in British Columbia, Canada, for his PhD. The data came from the start and stop location of each trawl tow over several years, a period that spanned the change in management from an "Olympic system" race to fish, to an individual transferable quota system where each fishing boat had a fixed quota for each species. What he found was that the trawl effort was very concentrated,[3] that is, the same areas were fished over and over again. Each boat had a set of areas known to be safe (no rocks), and with GPS the net could be set down along this same line each time. This concentrated effort meant that the actual total area trawled was much smaller than that reported using large statistical areas because some smaller areas were trawled many times each year.

There are now quite a few studies of trawl footprint that use even higher resolution data. The European Commission initiated a study of trawl footprint around the various EU regions and reported the trawl footprint for more than fourteen areas of the North Atlantic and Mediterranean waters using VMS data at a resolution of 1 minute of latitude and longitude (roughly 1 square nautical mile).[4] It found that the intensity of trawling was highly concentrated: 90 percent of trawling happened in 44 percent of the area, meaning that over half of the area was either lightly trawled or not trawled at all. In other words, with a fine-grained look, we discovered nuances in areas trawled and not trawled that the former coarse-grained investigations could not see.

Some areas are trawled more than ten times per year, whereas many areas are not trawled at all.

The results from this worldwide project, using high-resolution data from twenty-four regions and five continents, showed us great variations in intensity of trawling (see Figure 10.1 for examples).[5] Whereas the Europeans trawl the same ground over and over again, we saw that in eighteen of the twenty-four regions, two-thirds of the bottom had not been trawled in the 3 years covered by the available data.

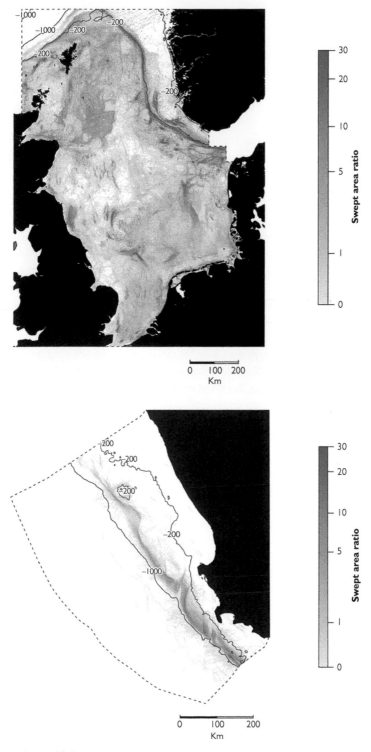

Figure 10.1 Trawl footprints for the North Sea and the Benguela Current off the West Coast of South Africa. Figures courtesy of Ricardo Amoroso.

The salient words here are *high resolution*. Trawl footprint data look very different depending on how large or small, i.e., coarse or fine, the units of area measurements are. There is a very strong relationship between data and size of the unit square. In other words, if you find trawling in a big square, the likelihood that only a portion of that square is actually being trawled is quite high. But the data do not reflect this. Any trawling activity in a big square suggests that the whole square is being trawled. However, if you break this big square into many smaller ones, several squares will turn out to be not trawled at all. The data will be concentrated in the squares that are actually being trawled.

A Case in Point

The TBP study estimated trawl footprint on 1 square kilometer of ocean and assumed that, within each square of such small size, trawling was uniformly distributed. In the shallow waters (under 200 meters) of the North Benguela Current off the Namibia coast, only a few percent of the 1 kilometers squares were trawled. But, if you increase the size of your squares up to 100 square kilometers of ocean, the same data will show that half of your big squares have been trawled. Since previous studies generally used much coarser scales, they grossly overestimated what proportion of the seafloor had actually been trawled. The 2002 study in the United States used squares of a minimum 25 square kilometers and up to over 1,000 square kilometers. Put trawl data into that kind of coarse resolution, and bingo! It looks like most of the ocean floor is being trawled, and the image of clearcutting the oceans could easily take hold in the public's mind.

Given That Quite a Few Areas Are Intensively Trawled, How Much of the Ocean Floor Is Really Being Trawled?

The 24 TBP areas with high-resolution data do not include much of the world's continental shelves. We have almost no high-resolution data from Asia, where trawling is often very intense. 20 percent of the global continental shelves are in the polar regions, where trawling hardly occurs. However, for the areas with high-resolution data, we did find that the proportion of the bottom trawled depended closely on trawl effort. In the seas around Europe, trawl effort is very high, and 50 to 80 percent of the bottom is trawled each year. In Australasia and the Americas, trawl effort is lower, and often less than 20 percent of the bottom is trawled. Using overall trawl effort data for the rest of the world, the TBP project was able to estimate that, globally, the proportion of the bottom swept annually by trawls is roughly only 15 percent.

What Does "Only 15 Percent" Mean When Compared with the Amazon Rain Forest?

According to TBP estimates, the total area swept by trawls each year is 25 million square kilometers, almost five times the size of the Amazon Rainforest. The vastness of the Amazon Rainforest we imagine is nothing compared with that of the oceans. The total continental shelf area is 32 million square kilometers, and over half of it is in Russia,

Canada, Australia, the United States, and the Antarctic where there is very little trawl-ing. But most of the trawled 25 million square kilometers are the same areas swept over and over again, so the actual fraction of the bottom being trawled is much less.

The big picture is that trawling does not affect most of the world's continental shelves, but it really hits some areas very hard. In those places, the *benthos* (com-prising the plants and animals on the ocean floor) is dramatically changed.

Impact of Trawling on Specific Biota

Once we know the footprint of trawling, we need to ask what passing a trawl net does to different species. Not surprisingly, it depends on (1) the species, (2) the type of sea floor, and (3) the type of trawl gear.[6] Species that are buried in the sediment, like clams and polychaete worms, are pretty resistant to many kinds of trawl gear. Species that sit on the surface and rise up like corals and sponges are much more likely to be killed or damaged by a passing net. The type of ocean floor will affect the same species differently. In general, hard sea floor and gravel tend to be the most disturbed, whereas mud and sand are less so, partly because what the sediment looks like frequently indicates how often that particular ground has been subject to natural disturbances. Finally, the type of net makes a big difference. An otter trawl only digs into the sea floor where the doors touch it. The rest of the gear tends to ride on the surface. Beam trawls drag a very heavy piece of metal, and while it does not usually dig into the sediment, it is going to knock down almost any structure that rises above the seafloor. Scallop dredges, whose teeth dig into the sediment, will have by far the biggest impact.

There are two kinds of studies that look at the impact of bottom trawl gear. One is to compare sites that have been trawled to those that have not—these are called compara-tive studies. The second type of study is experimental, comparing the biota before and after a trawl has passed over the bottom. The TBP project found twenty-four compara-tive and forty-six experimental studies on the impact of trawling on benthic biota. Otter trawls had the least impact: they removed an average of 6 percent of organisms per pass. Beam trawls removed 14 percent per pass, and scallop dredges 20 percent. The impacts are closely related to how far the gear penetrates the ocean floor.

We can now estimate the fraction of different types of animals killed when a net passes over different types of habitat (see Table 10.1).

Table 10.1 *Fraction of Three Different Species Killed by One Passage of an Otter Trawl over Different Habitats*

	Mud	Muddy-sand	Sand	Gravel
Bristle worms	0.27	0.41	0.38	0.49
Crabs shrimp isopods	0.59	0.66	0.65	0.71
Clams	0.02	0.21	0.16	0.31

Net Impact of Trawl Footprint and Impact of Trawling

The net impact of trawling on a certain species in a defined location depends on (1) how many are killed or damaged by the passage of a net, (2) how often the same location is trawled, and (3) how fast the affected species recovers. Recovery is closely related to lifespan—long-lived species like corals and sponges may take decades to recover, while short-lived species can recover quite rapidly. The TBP group estimated recovery rates by examining how much the relative abundance of a species depends on how often an area is being trawled. Combining the three factors allowed us to estimate the total impact on different taxa for different places, and we can then map the distribution of individual organisms, or total benthic abundance, for our twenty-four comparative studies.

Here Is What a Lightly Trawled Area Looks Like

Figure 10.2 is typical of a reasonably lightly trawled location. Over most of the area there is little trawling and no impact on the benthos, while most of the trawl effort is concentrated in certain locations, and there the benthos is much reduced. Anything long-lived will be gone, and only the most resilient and rapidly recovering species survive. The important message from these data is that fishing effort is very consistently concentrated in specific locations. This might come as a surprise to anyone but the fishers—they fish in the same places year after year because that is where the fish are.

The habitat relative benthic status (RBS) is the abundance of benthic biota compared with what it would be in the absence of trawling. Some areas show almost total loss of benthos, but most areas show no impact.

So What about Clearcutting in Heavily Trawled Areas?

In heavily trawled areas, the benthic community has been greatly reduced. Any sea floor structure of organisms that used to reach up above the floor will have been flattened by the repeated passages of trawl nets, and physical structures of rocks and gullies will have been simplified. Such structures, whether biological or geological, are often part of the habitat for target species. Cracks and rocks may provide protection from predators for small fish and invertebrates, but at the same time, some of the benthic animals that have been killed may be food for some target species. Just the same, one might be forgiven for thinking that intense trawling will indeed reduce the productivity of target species that depend on the structure for their survival.

But hold on—could it be that stirring up the bottom sediments makes the ecosystem more productive and brings to the surface both nutrients and food that would otherwise be hard to find? So—perish the thought—perhaps trawling actually benefits some target species at the same time it diminishes other species that depend on the structure.

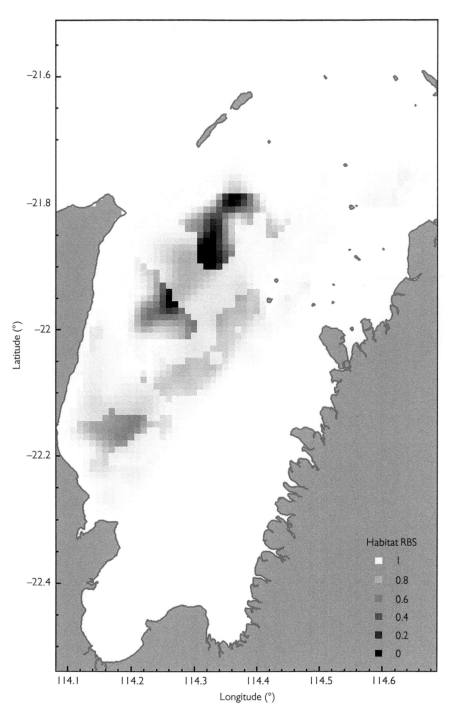

Figure 10.2 Impact on benthic biota from trawling in Exmouth Gulf, Australia. Figure courtesy of Roland Pitcher. Redrawn from Pitcher C R, Ellis N, Jennings S, Hiddink J G, Mazor T, Kaiser M J, Kangas M I, McConnaughey R A, Parma, A M, Rijnsdorp A D, 2017. Estimating the sustainability of towed fishing-gear impacts on seabed habitats: A simple quantitative risk assessment method applicable to data-limited fisheries. Methods in Ecology and Evolution 8: 472–80.

If we believe the anti-trawling web sites, trawling would appear to destroy ecosystems and everything in them. Accordingly, a heavily trawled area should be a desert without benthos, without fish. Really? So how can it be that in areas where we have good data, the fishing boats go to the same places year in and year out, and the target species are still there and get caught without any reduction in tonnage? So much for clearcutting and creating a desert.

The North Sea now has a swept area ratio of 1.2, considerably less intense than it used to be. This means that the same square meters are trawled, on average, more than once a year. If trawling truly destroyed ecosystems, we could no longer expect to find benthic species like cod and haddock, or at the very least, they would have to be less productive than they were in the past.

Far from it, cod and haddock are still productive and abundant in the North Sea. Even more contrarily, during the 1970s, there was a boom in cod and haddock, called *the gadoid outburst*, gadoid being the taxonomic group that encompasses both species. Despite having been trawled heavily for a century, the North Sea managed to spring a surprise outburst on everyone.

In sum, the TBP[7] project examined whether trawling indirectly reduced or increased the abundance of target species by modifying the benthic community. We found some evidence for both effects, but it was always weak. Bottom trawling simply does not appear to strongly impact the target species other than by catching them.

Best Practices

What Would a Best Practice for Managing Trawl Fisheries Look Like? And If a Project Like TBP Were to Recommend One, Would It Be Applicable Everywhere?

Clearly, different countries have different priorities when they weigh food and jobs against protecting specific benthic organisms. For the United States, protecting corals and sponges will weigh heavily, whereas for Vietnam or Mozambique, food and jobs may be more important. With this in mind, the TBP chose not to recommend specific best practices, but rather to lay out what they are and what the costs and benefits of implementing them would be.

It became obvious early on that the big concern was with long-lived species, such as corals, sponges and sea whips, that reach above the sea floor and are very vulnerable to trawl gear. Burrowing worms, clams, and small invertebrates were not on anyone's must-protect list. Since they are the long-lived species that are most affected by trawl gear, any best practice must focus on them.

The major possible best practices include:

- *Prohibition of specific gears.* Some trawl gears have been designed to fish on the rough sea floor where sensitive species are found. Also called "roller gear," truck tires are attached to the bottom of the net, allowing the gear to roll over rough objects. Banning such gear would make fishing the hard rocky sea floor much more difficult.
- *Modification of gears.* Fishing gear can be modified in a wide range of ways to reduce benthic impact. Given that otter trawls are the most common gear and that the doors are extremely destructive, modifications have been designed to reduce or in some cases, even eliminate contact with the sea floor.
- *Reducing trawl effort.* This is the simplest way to reduce benthic impacts, and it has been applied in most of the developed world over the last 10 years. A total ban, while really good for the benthos, carries a very high price in terms of lost catch.
- *Freezing the footprint of trawling.* Wherever vessel locations can be tracked, the current trawl footprint could be frozen. Fishing boats would be restricted to within the bounds of areas already being trawled. This would make the fishery less resilient to climate change, which will likely change where the fish can be found.
- *Closing specific areas.* This is the currently preferred best practice. Sensitive habitats like seagrass beds are closed, and wherever coral and sponge habitat is mapped, closure would be beneficial. Many countries also prohibit trawling close to shore to protect small-scale inshore fisheries.

Each of these options is used somewhere in the world. But we must always remember that implementing each one of them may carry a hefty price in loss of catch or economic hardship for boats, crew, and their communities.

FURTHER READING

Typical NGO site vilifying trawling. Greenpeace 2018. Trawling destroys the seabed. https://www.greenpeace.org.uk/what-we-do/oceans/overfishing/bottom-trawling/.

The paper suggesting trawling is like Amazon forest clearcutting. Watling L and Norse E A. 1998. Disturbance of the seabed by mobile fishing gear: A comparison to forest clearcutting. Conservation Biology 12: 1180–97.

A major independent review of the impacts of trawling. NRC 2002. Effects of trawling and dredging on seafloor habitat. Washington, DC: National Academy Press.

A European study of trawl footprint. Eigaard O R, Bastardie F, Hintzen N T, Buhl-Mortensen L, Buhl-Mortensen P, Catarino R, Dinesen G E, et al. 2016. The footprint of bottom trawling in European waters: distribution, intensity, and seabed integrity. ICES Journal of Marine Science 74: 847–65.

The Trawling Best Practice estimate of footprint. Amoroso RO, Pitcher CR, Rijnsdorp AD, McConnaughey RA, Parma AM, Suuronen P, Eigaard OR, Bastardie F, Hintzen NT, Althaus F, Baird SJ. 2018. Bottom trawl fishing footprints on the world's continental shelves. Proceedings of the National Academy of Sciences 115(43): E10275-82.

Estimation of the impact of changing benthic biota on target species productivity. Collie J, Hiddink J G, Kooten T, Rijnsdorp A D, Kaiser M J, Jenning, S, and Hilborn R. 2017. Indirect effects of bottom fishing on the productivity of marine fish. Fish and Fisheries 18: 619–37.

Estimates of passage of different types of trawl gear on different biota. Hiddink J, Jennings S, Sciberras M, Szostek C L, Hughes K M, Ellis N, Rijnsdorp AD, McConnaughey R A, Mazor T, Hilborn R, Collie J S, Pitcher R, Amoroso R O, Parma A M, Suuronen P, and Kaiser M J. 2017. Depletion and recovery of seabed biota following bottom trawling disturbance. Proceedings of the National Academy of Sciences July 17: 201618858.

Integrating trawl footprint with mortality of different gear to estimate net impact on biota. Pitcher C R, Ellis N, Jennings S, Hiddink J G, Mazor T, Kaiser M J, Kangas M I, et al. 2017. Estimating the sustainability of towed fishing-gear impacts on seabed habitats: A simple quantitative risk assessment method applicable to data-limited fisheries. Methods in Ecology and Evolution 8: 472–80.

CHAPTER 11

The Forage Fish Rollercoaster

John Steinbeck + Cannery Row = Sardines

Steinbeck's *Cannery Row* was written when the California sardine fishery was booming; the depression bit hard, but the sardines were doing pretty well. The famous canneries are no more, but one of them has been glamorously converted into the Monterey Bay Aquarium, where you can admire the shimmering, silvery sardines in their iconic, tightly packed schools in a vertical, tubular tank, as they circle, switch direction, flashing their silver, and turn again, a mass of dark, small forms. Then you can watch an equally impressive silvery, tightly packed school of anchovies in another tubular tank and despair over your inability to memorize the differences between one small fish and another.

Fishing for sardines in California began in earnest in the early twentieth century, when modern canning equipment became available and purse seines were perfected to efficiently catch schools of fish. In the cannery museum, part of the Aquarium building, you can see wonderful old promotional films of the process of canning fish from the arrival of the boats at the piers to labeling and shipping the cans all over the United States.

By the 1920s, much of the sardine catch was processed into fishmeal and oil for the poultry industry. (Ulrike speaking: One of the first of many startling and inexplicable impressions after my arrival in the United States in the early 1960s was the decidedly fishy taste of chicken. It put me off chicken for years.) In the 1940s and 1950s, the sardine fishery boomed. Annual catches reached 500,000 tons, equivalent to 4 million head of beef. The decline came in the 1960s (see Figure 11.1). Canneries closed; fishing boats sat idle; and families who had fished sardines for generations struggled to find alternative ways of life. The final blow came in 1968, when the sardine population went down to only a few thousand tons of total abundance, several hundred times smaller than it had been, and the fishery closed.

In time, the population built up again. In the 1990s and in the 2000s, it reached 500,000 to 900,000 tons. Management handled this boom much more cautiously

Ocean Recovery: a sustainable future for global fisheries? Ray Hilborn and Ulrike Hilborn, Oxford University Press (2019). © Ray Hilborn and Ulrike Hilborn 2019. DOI: 10.1093/oso/9780198839767.001.0001

so that average catches came in at only 80,000 tons. Nevertheless, the sardine collapsed once more, and the fishery was closed again in 2016.

What to Make of It All?

In the 1960s and 1970s, the California sardine was considered a classic example of overfishing that found its way into many fisheries textbooks. However, subsequent research has shown that sardines rise and fall steadily over centuries. In view of these natural population cycles, the 1950s collapse is now recognized as inevitable but hastened by overfishing, while the 2010s collapse took its natural course without being speeded up by the fishery.

What to Do with All the Idle Boats and Processing Plants?

When the sardine fishery began collapsing, fishing companies looked for a new fishery suitable for producing fishmeal and fish oil, and found it with the large

A Closer Look

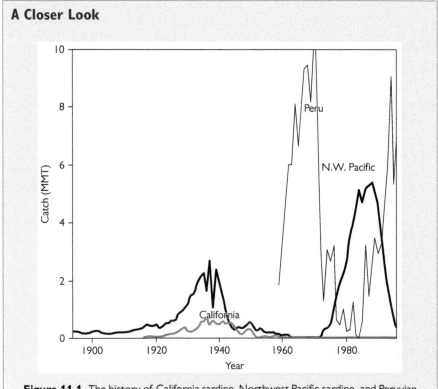

Figure 11.1 The history of California sardine, Northwest Pacific sardine, and Peruvian Anchoveta catches have all shown boom and bust in catch reflecting periodic increases in productivity. Redrawn from.[1]

anchoveta populations off Chile and Peru. Many of the larger boats were moved south, and more modern reduction plants were built there.

The Peruvian anchoveta fishery boomed in the 1960s, at over 10 million tons, twenty times more than California at its height, collapsed in the early 1970s, stayed at low abundance through the 1980s, and then built up again.[1]

The sardine fishery in Japan that is ten times larger than the California sardine fishery underwent similar fluctuations with a boom in the 1930s and 1940s, a collapse in the 1960s and 1970s, and a boom again in the 1980s and 1990s.

What Makes a Fish a Forage Fish?

Sardines and anchoveta are members of a super important but underappreciated group of fish called forage fish. They are the small fish that eat zooplankton and other small aquatic organisms that form the base of the marine fish food chain, also called baitfish because they generally are food for the larger, prized game fish. Sardines, mackerel, herring, menhaden, and anchovies are the most numerous forage fish, but there are also shad, capelin, and smelt.

For many reasons, forage fish are socially, economically, and biologically important, particularly, because they are the most abundant fish we harvest and they form very large schools that are easily captured, usually by surrounding them with a net. For cost and fuel used per ton landed, they are among the most efficient capture fisheries in the world and perhaps have the lowest greenhouse gas footprint of any way to produce protein. They are also the largest fisheries in the world and vital to the food security and economy of many countries. But, they are also highly variable: many have shown one hundred-fold fluctuations in abundance over decades, which poses very special challenges to sustainable management. Finally, since they are the primary food for many marine fish, mammals, and birds, there are trade-offs between harvesting forage fish and consequences for their predators.

How Forage Fish Are Used

(1) As food for you and me, either fresh (grilled sardines!), frozen, canned, smoked, or salted.
(2) For fishmeal and fish oil.

The latter is used primarily for animal feed, but increasingly as a dietary supplement for people. Forage fish have become the dominant source for fishmeal and fish oil, and both have become essential to the aquaculture industry. About 70 percent of modern aquaculture raises piscivorous fish, that is fish that eat other fish. Unlike farm animals, piscivorous fish cannot turn carbohydrates into meat: they need protein. Even though some plant protein can be fed in aquaculture, fishmeal and oil are vital to the industry.

Lately, using forage fish for fishmeal and oil has become controversial. To use relatively low value fish to produce high value salmon for rich consumers is seen by some as a threat to food security. To others, the inefficiency of using lots of small fish to produce bigger fish is indefensible. The Aquaculture Stewardship Council uses a "forage fish dependency ratio" as a measure of sustainability, suggesting that using lots of forage fish is less sustainable than using plant feed.

Fishmeal and fish oil producers contend that quite often there is almost no demand for forage fish as human food because they taste—well, too fishy. They also point out that in modern aquaculture more than a kilo of fish meat can be produced with much less than a kilo of forage fish because of the highly efficient feed conversion by aquaculture fish and the use of plant protein as diet supplements. They also note that with modern processing an increasing proportion of fishmeal and oil consists of formerly discarded byproducts from fish destined for human consumption.

The Puzzle of the Rollercoaster: How Do We Even Know It Exists?

Boom and bust is what forage fish do quite independently of both fishing and management. These fluctuations come in long enough periods of somewhere between 10 to 100 years, and intensive commercial fishing is barely 100 years old, so when the calamitous collapse happened in the 1960s, scientists felt justified to blame overfishing.

In Europe, where fishing for forage fish has a much longer history than in California, the boom and bust cycle had been noticed long before industrial-scale fishing. David Cushing, in *"Climate and Fisheries,"*[2] describes the history of herring fishing in Sweden and Norway. Over the centuries, herring continuously went up and down—when fishing was good in Sweden, it was typically poor in Norway.

Of Cores and Fish Scales

Lots of science has gone into proving that forage fish cycles go back to prehistory and have not been influenced by fishing.

Very conveniently to this task, there are small oceanic dead zones in places like the Santa Barbara basin in southern California where fish scales are well preserved because the lack of oxygen means the scales do not decompose. In other words, fish bones and scales do not rot. Year after year, scales and fine dirt from the rivers are deposited for scientists to read much like a history book.[3] This book is lifted from the ocean bottom via hollow metal cylinders that are pushed into the sediments. But, instead of pages, up comes a core, a striped noodle of sediment and scales that tells a story of ever-recurring scarcity and abundance, and annual layers of silt from winter storm runoff that mark the passing of the years (see Figure 11.2).

You will have to forgive this simplistic presentation of highly complicated historic reconstruction, years of field work on the water drilling cores, arduous staring into microscopes and identifying which kind of fish dropped what scales when, research and analysis, and fitting it all into the framework of prehistory before telling the world about it in scientific journals.

A Closer Look

These graphs show a nearly 2000-year reconstruction of the abundance of sardine and anchovy off the coast of California based on scale deposition rates in anoxic sediments. This shows the very wide fluctuations seen prior to any industrial fishing.

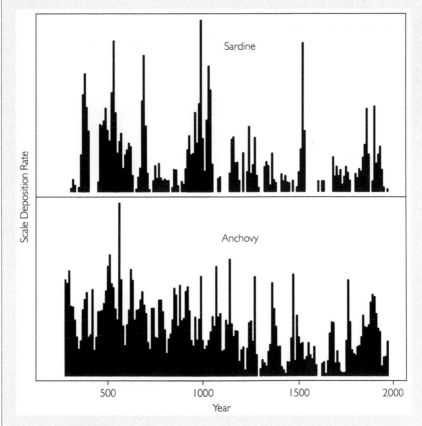

Figure 11.2 Estimated abundance of sardine and anchovy off the California coast for the Last 2000 Years.

Redrawn from Schwartzlose RA, Alheit J, Bakun A, et al. 1999. Worldwide large-scale fluctuations of sardine and anchovy populations. S Afr J Mar Sci 21: 289–347.

Why Rollercoaster Forage Fish Are a Real Pain for Managers

How do we deal with the concept of maximum sustainable yield when confronted with wildly fluctuating abundance that is out of our control? The conventional view of fish population dynamics is that, in the absence of fishing, the stock will fluctuate around a "carrying capacity" and that long-term harvest is maximized by fishing hard enough to reduce that average abundance to roughly one-third to one-half of the unfished abundance. Stocks that fluctuate ten-fold in decades long cycles simply do not fit that idea of sustainable fishing.

There are certainly theories on why forage fish oscillate so inconveniently. Is it changes in the environment? Is it the abundance of predators and prey, or competitors, or physical changes such as ocean temperature? Take your pick. What is clear is that for most forage fish, there is little if any relationship between how many spawn and how many offspring survive to maturity. This is counter intuitive—if twice as many eggs are put into the water, a reasonable assumption would be that twice as many fish will survive to maturity in good and bad conditions— but not so. The data show almost no relationship when the conditions are accounted for.

Never Mind the Managers, What about the Fishermen?

Fishing communities and fishermen bear the brunt of the forage fish's unreasonable behavior. You cannot specialize in a wildly fluctuating fishery and stay in business—yes, there are decades of good times when the fish run high and money is made, but when the stocks dive, those good times are over. When the California sardine tanked, some of the fishermen and processors could move to new fisheries in Chile and Peru, but the majority of fishermen could not and either went out of business or diversified into other fisheries. That was then. Now, in modern fisheries, that kind of switching to new countries or other fisheries is difficult if not impossible.

No new major fisheries have been found recently, and the widespread use of limited entry usually prohibits a sardine specialist to switch to something else. The California sardine fleet happily has the option of fishing for squid and that has become the most valuable fishery in the region.

The Boom and Bust Is Not Only Temporal but also Spatial

During a boom, there are so many fish that they need more space, and the resulting range expansions are quite spectacular. In years of super abundance, California sardines reach up into Canada, and the Chilean jack mackerel population extended all the way across the Pacific at peak abundance, creating new headaches for managers. New countries and new fishing communities suddenly have access to this bounty, and the hard work of sharing out the fish begins.

Present Status and Management of Forage Fish Stocks

As we would expect of rollercoaster stocks, at any one time, some will be at high abundance, some at low. Let us compare the most recent assessments to the highest abundance ever recorded.[4] The average ratio of now to best ever is 0.46; stocks are on average about half of the highest high, and that is what we would expect of randomly fluctuating stocks.

On the low end, the Japanese sardine is presently at 4 percent of its high of 10 million tons, and the Chilean jack mackerel is at 20 percent of its high of 23 million tons. On the high end, a few stocks are at or near record highs: Northeast Atlantic mackerel, Gulf of Mexico menhaden, Bothnian Sea herring, and Chilean herring. Most of the stocks, though, are somewhere in between.

Are We Starving Natural Predators by Fishing?

Given the forage fish's low rank on the food chain, it would seem obvious that by fishing we are taking food out of the mouths of their natural predators. Several NGOs have spoken up in favor of adopting a very precautionary management of forage fish and stopping any development of new forage fish fisheries.

But, in fisheries, nothing is that simple.

Any predator specializing in a boom and bust forage fish would not last the distance. There is a whole ecosystem offering up alternatives. Evolutionarily speaking, predators have learned to diversify to a mixed diet.

Of concern to NGOs were, for instance, pelicans and sea lions that feed on California sardines. Since in most ecosystems no individual prey species constitutes more than 20 percent of potential edibles, pelicans, and sea lions have long survived on a mixed diet. To profit most from the sardines, though, pelicans choose their breeding sites carefully. Because even in times of scarcity there will be some sardines, pelicans breed where they are reliably found.

Meanwhile, the alternative forage species seem to have doubled in abundance globally, most likely because their natural predators have been harvested and, happy days, their own competitors, the commercially important forage species, have met the same fate.

About Fish-Eating Guano Birds Whose Poop Once Turned into Fertilizer Gold

We are not talking about the guano gold rush of the nineteenth century, but about the birds themselves, whose poop once created true white mountains of momentous, well—you know what. The Guanay cormorant, principal creator of the guano mountains of yore, the Peruvian booby, that makes nests almost exclusively of its

A Closer Look

Pikitch et al. 2012[5] used ecosystem models to evaluate the impact of fishing forage fish on predators and concluded that there was a strong impact to be expected and that forage fisheries needed to be managed very conservatively. The models they used were very unrealistic. They did not include any natural variability in forage fish (perhaps the most obvious characteristic of forage fish populations), they assumed a strong relationship between forage fish spawning stock and subsequent recruitment, they did not consider that the size of fish taken by predators might be different from the size of fish taken by the fishery, nor did they consider any spatial variation in distribution as abundance fluctuated.

Subsequently, Punt et al. 2016[6] looked at the California sea lion and pelican interaction with sardines and anchovy with a model that included all the elements that Pikitch et al. did not use. They found almost no impact of fishing on the pelicans, and none on the sea lions (see Table 11.1).

Table 11.1 *Changes to Pelican Populations Attributable to Fishing as Estimated by Punt et al. 2016*

Species and policy	Average population size	Proportion of time < 0.5 K	Proportion of time < 0.1 K
Pelican without fishing	0.942	0.042	0.009
Pelican with fishing	0.925	0.053	0.011

own poop, and pelicans, stalwart contributors, all feast on the Peruvian anchoveta, the largest of all forage fish stocks with a fishery to match.

Because of the size of the anchoveta fishery, the question has arisen whether the guano birds suffer from a reduced supply of food.

The birds certainly declined sharply when the fishery first developed in the 1950s and 1960s, but since then, there is no apparent relationship between anchoveta abundance and bird numbers. The big question is why the birds declined in the 1950s and 1960s when anchoveta were very abundant. It has been suggested that abundance alone is not what matters, but whether the birds can actually catch the fish when the fishery breaks up large concentrations of anchoveta near the bird colonies. A recent paper found some evidence of this for boobies and cormorants, but not for pelicans.

There is of course always an exception, even to the forage fish oscillations, and that would be the menhaden of the Atlantic and Gulf of Mexico, two large US forage fisheries whose catch is almost exclusively processed into fishmeal and oil. Menhaden do not appear to show (at least in recent decades) any dramatic ups and downs. No rollercoaster for them.

A Closer Look

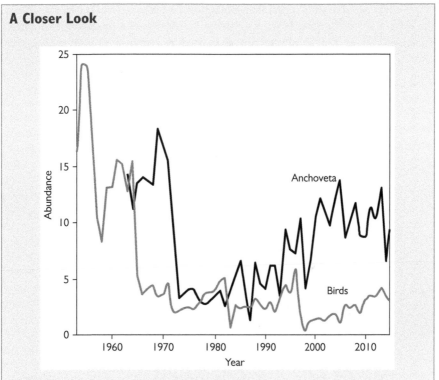

Figure 11.3 The history of number of guano birds (millions of individuals) and anchoveta (millions of tons from survey data) with a major decline in bird abundance taking place in the 1960s while the anchoveta was still at high abundance. Since the 1990s, anchoveta abundance has increased considerably, but birds have remained at low abundance. Data IMARPE—Peruvian Government Fisheries Research Agency.

Here the concern is not food for birds, but food for growth and survival of some valued recreational and commercial fisheries, particularly striped bass.

We have good estimates of the abundance of Atlantic menhaden over 30 years, and they have fluctuated five-fold. The scientists specializing in menhaden have found no relationship between how many spawned and how many reached maturity—if anything, there have been slightly fewer reaching maturity at high spawning abundance. Thus fishing does not affect the number born or surviving their first year, but only how many reach older ages. The fishery targets 2 year olds and catches very few 1-year-old fish. Striped bass, however, mostly eat younger menhaden before they are two years old. Therefore, the fishery is not taking food out of the bass's mouths.

It does, however, potentially take food out of the mouths of bluefin tunas, who prefer their menhaden to be 2 and older.

All our analyses have failed to show any relationship between abundance of forage fish and growth, or decline of their predators. Brown pelicans and sea lions in California increased rapidly when sardines were almost absent.

We conclude that each individual problem in the complex interactions between the fishery, forage fish, and their predators merits its own scientific investigation. There are no simple general rules, and expectations of the seemingly obvious have often turned out to be wrong. Finally, we also need to consider the social implications of assigning value before we make management decisions. What value do we as a society assign preserving the highest number of predators versus protecting human food supply and jobs?

• •

REFERENCES AND FURTHER READING

Baumgartner T R, Soutar A, and Ferreira-Bartrina V. 1992. Reconstruction of the history of Pacific sardine and Northern anchovy populations over the past two millennia from sediments of the Santa Barbara basin, California. CalCOFI Report 33: 24–40.

Pikitch E, Boersma P D, Boyd I, Conover D, Cury P, Essington T, Heppell S, et al. 2012. Little fish, big impact: managing a crucial link in ocean food webs. Washington, DC: Lenfest Ocean Program, 108.

Punt A, MacCall A D, Essington T, Francis T B, Hurtado-Ferro F, Johnson K F, Kaplan I C, Koehn L E, Levin P S, and Sydeman W J. 2016. Exploring the implications of the harvest control rule for Pacific sardine, accounting for predator dynamics: A MICE model. Ecological Modelling 337: 79–95.

Schwartzlose R. and Alheit J. 1999. Worldwide large-scale fluctuations of sardine and anchovy populations. South African Journal of Marine Science 21: 289–347.

CHAPTER 12

Following the Rules and Illegal Fishing

Following the Rules: The Key to Effective Fisheries Management

What is the point of rules and management if they are not followed? Whether it is a small fishing community that has agreed on rules or industrial fisheries on the high seas, if the rules are broken, they are meaningless. In our survey of the global status of fish stocks, compliance with rules really matters.[1]

Freedom of the Seas

Nowhere on land is there a fantasy of absolute freedom as on the high seas. Escape the drudgery of ordinary life to spend your time sailing across the oceans, unfettered by responsibility or regulation is many a person's idea of "freedom of the seas." It is even a legal doctrine. It is not too long ago that there were no rules and no police on the high seas; once you were beyond territorial waters, you were your own master.

Well, not quite. By agreement, when in international waters, vessels must be registered in an individual country and obey its regulations that mostly cover safety, inspections, and wage standards. That would, of course, be nice, but one can avoid such onerous standards by registering one's vessel in countries with loose standards (some of them even landlocked), where registration is cheap, and safety and inspection requirements are risible.

Enshrined in the UN Convention on the Law of the Sea in its 1994 version, are specific freedoms that include navigation, laying submarine cables, scientific research, and fishing. Unless otherwise restricted by international treaties, any high seas vessel is free to catch as much as it wants in any way it wants, and could

Ocean Recovery: a sustainable future for global fisheries? Ray Hilborn and Ulrike Hilborn, Oxford University Press (2019). © Ray Hilborn and Ulrike Hilborn 2019. DOI: 10.1093/oso/9780198839767.001.0001

kill birds, mammals, and turtles, and could and did drive many whales to near extinction. On the high seas, you could once, indeed, do as you pleased.

The Constraints to Freedom of the Sea: Best Intentions, Small Effects

To bring this lawlessness under some sort of control, all manner of organizations with all manner of goals were established, such as the International Maritime Consultative Organization in 1948, with a name change to International Maritime Organization in 1982, that led to the adoption of several international treaties primarily around safety at sea and prevention of marine pollution. One of those was the 1972 Convention on the Prevention of Marine Pollution by Dumping of Wastes and Other Matter, whose effectiveness may be a matter of degree only.

Concerning fisheries, the need for international cooperation became obvious much earlier. There were the North Sea Fisheries Convention of 1882, the International Council for the Exploration of the Sea in 1902, the International Pacific Halibut Commission in 1923, the International Whaling Commission in 1946, the Inter-American Tropical Tuna Commission in 1949, the Great Lakes Fisheries Convention in 1955, the Convention on Fishing and Conservation of Living Resources of the High Seas in 1966, the International Commission for the Conservation of Atlantic Tunas in 1969, and the Convention for the Conservation of Antarctic Living Marine Resources in 1982. There are now dozens of treaties and commissions around the world designed to better manage high seas fisheries and shared fish stocks.

The Law of the Sea: Never Ratified by the United States

The most significant agreements have been brokered by the United Nations, and these include the Law of the Sea (1994), an overall agreement that formalizes the 200-mile zones and sets guidelines for international fisheries and other maritime activities. Another one is the UN Straddling Fish Stocks Agreement[2] about highly migratory fish stocks and fisheries that are shared between countries.

Despite the best intentions, these treaties and the agencies they established are deeply flawed. Any bite they might have had simply was not there. Organizations were given authority for scientific research, and to make science-based management recommendations to member States. Any power of enforcement was usually given only to the flag State. And whatever decisions were made, bound only the signatory countries to comply, those who did not sign on were free to do as they pleased. Monitoring of compliance by the signatory countries was minimal. The United States, for example, never managed to sign the Law of the Sea.

The widespread adoption of the 200-mile exclusive economic zone should have brought management by adopting countries to their own fishing grounds where regulations could be enforced. Theoretically, however, many countries, despite having established a 200-mile zone, regulate little and enforce less. Unauthorized foreign vessels often freely fish in a nation's waters without fearing punishment, and even a country's own fishermen often thumb their noses with impunity at their own regulations.

Illegal Fishing

The Time Has Come to Open the Rat Bag of Illegal Fishing

"Illegal, unreported, and unregulated (IUU) fishing" is fishing that violates regulations or undermines conservation and management actions. The Food and Agriculture Organization of the United Nations (FAO) estimates that IUU catch could be in the range of 10 to 25 million tons per year,[3] possibly more than a quarter of all reported landings, and divides IUU fishing into four categories.

(1) Fishing and fishing-related activities in contravention of national, regional, and international laws.
(2) Non-reporting, misreporting, or underreporting information on fishing operations and their catches, otherwise called cheating.
(3) Fishing by "stateless" vessels.
(4) Fishing in designated areas of a Regional Fisheries Management Organizations (RFMOs) by out-of-region vessels.

We may think that IUU fishing is primarily a problem of developing countries that have neither the money, nor the coast guard or the know-how to enforce regulations when dealing with their own fishermen, or worse, keep foreign vessels out of their waters, but rich countries are not immune either. Abalone poaching is a persistent problem in New Zealand and Australia. High value and sought after, abalone are often sold to restaurants from the back of a truck or smuggled to Asian countries. In the Auckland airport, suspiciously long lines of baggage for check in by tour groups more than once revealed suitcases full of dried abalone.

Because its true extent is unknowable and guess work at best, IUU fishing creates problems for the scientific processes that should lead to sustainable management, and it constitutes outright theft of fish from the rightful owners.

High-Value Southern Bluefin Tuna

A case in point is the southern bluefin tuna, a high-value fishery. When the fish are young, they are mostly caught in Australia's economic zone, when older, in the Indian Ocean. The fishery is managed by the Commission for the Conservation of Southern Bluefin Tuna. Japan, Australia, and New Zealand founded it because it became apparent that the bluefin stock had reached appallingly low levels and would profit from catch reductions to rebuild. Starting in 1989, the commission agreed to strongly reduce catches. The scientific committee's computer models confidently predicted that the stock would rebuild—but it did not, it stayed in the basement. I joined the scientific committee in the mid-1990s and was a formal member from the late 1990s to the mid-2010s. There was great consternation among us.

Why, we wondered, instead of rebuilding, did the stock keep going down? By 2004, we had to recommend further quota cuts. With some sleuthing, the Australians discovered that the volume of southern bluefin sold in Japan was three times the reported catch.[4] Independent investigators agreed with the numbers, and, while no specific country was named culpable, Japan consented to a further reduction of its allowable catch. A range of additional enforcement measures were carried out, and the stock has been rebuilding ever since.

West Africa

West African countries also have a serious problem with IUU fishing. They rely on small-scale coastal fishing for most of their catch and cannot effectively protect their waters from being plundered by foreign vessels, be they authorized and cheating on the catch reports, or be they unauthorized and poaching inside the exclusion zone. There are suspicions that the IUU catches off West Africa are as much as 50 percent of the countries' reported landings.

A further complication is vessels that fish off West Africa with legal permits that may have been obtained through corruption. Technically, that is not poaching, but each country's people are still being cheated of the benefits that should have accrued them from their resource.

The Antarctic

In the vast waters of the Antarctic, poachers have long been busy catching valuable Patagonian toothfish, also called Chilean sea bass. In the 1990s, it was estimated that the IUU catch was up to six times the legal catch. Illegal vessels do not just poach fish: they are not exactly keen to protect albatross and whatever else they catch incidentally. Despite a downturn in illegal catches overall, there are probably still twenty to thirty ships fishing for toothfish illegally and raking in the money.

The Antarctic division of the Australian Government has a long and enlightening list of the sophisticated weaving and dodging by the illegal vessel networks:

- Sophisticated control of vessel movements.
- Complex logistics, including chartering tankers to refuel vessels at sea, specially built and modified vessels, and the use of ports in States that give little scrutiny to vessel movements and catch landings.
- Use of active intelligence gathering about States' enforcement efforts.

- Use of complex corporate arrangements and legal advice to exploit weaknesses in national and international fisheries and corporate law and to disguise the real owners and beneficiaries of IUU fishing.
- Use of weak State regulatory regimes to flag vessels and fraudulently obtain validation of catch documents needed for market access.
- Use of complex arrangements to "launder" catches.
- Strong evidence of corrupt conduct in support of IUU fishing by some countries' officials.
- A small number of beneficiaries controlling nearly all IUU fishing for toothfish.
- Substantial and well-financed legal challenges when vessels are arrested.

Enforcement: How Is It Done?

Enforcement of fisheries regulations consists of monitoring, control, and surveillance (MCS). All MCS systems depend on a legal framework and enforceable penalties to be effective. Techniques used can be those of traditional law enforcement, including informants and undercover officers, or a number of well-known standard methods.

The most basic of these are catch and location records that vessels are required to keep and that are compared with inspections done at the time of landing. Of course, a vessel can cheat by landing at remote locations that are not monitored, transferring the catch at sea to vessels destined overseas, and hiding portions of the catch in secret compartments.

Vessel Monitoring Systems and On-Board Observers

There are now vessel monitoring systems (VMS) that track a ship's identity and location by sending a signal to a satellite, roughly about every 2 to 4 hours. They have become standard on larger vessels around the world, including many from developing countries—but they can be prohibitively expensive for smaller vessels and impractical for artisanal ones. Cheaper technology is now available but not yet widely used. Like a smartphone, these small, cell phone sized units are equipped with a GPS system and record a ship's location almost continuously. The data are uploaded via cell phone networks once the ship returns to port. It is quite possible that in the future, all small-scale fleets and even artisanal fisheries will be equipped with these devices.

Vessel tracking can be used to enforce closed areas and spatial allocation of effort. For instance, if a vessel reports catch from a certain area and the VMS data show the vessel was not there, the vessel can be charged with illegal fishing. Fine scale tracking from near continuous GPS recording can distinguish between fishing methods, whether a boat was longlining, purse seining, or trawling, and whether it was in transit or actively fishing.

Some countries use patrol flights over their economic zones to look for unauthorized vessels. This is expensive, but it provides some of the best estimates of how often such ships come into their waters. In recent years, the use of drones has been promoted to replace the more expensive airplanes. Illegal vessels can also be spotted and reported by legal fishing vessels, and, if the coastal country has the resources, a patrol vessel or aircraft can try to arrest or at least identify the offender.

No VMS can tell us how many fish are caught, what fish are being discarded, and if any birds, marine mammals, or turtles are trapped by a net or caught on a hook. On-board observers (or electronic monitoring systems, see the next section) are needed to detect bycatch and discards. Many countries use them, sometimes purely for scientific data collection, but some others for enforcement as well. The observers record locations of active fishing, what is caught, and what is discarded. Sometimes, they will also take scientific samples of species composition and size and perhaps take samples for genetic analyses or to age the fish. In Alaska, all larger vessels are required to have an observer on-board, and vessels that process their catch at sea are required to have two, so that one observer is always on duty. British Columbia bottom trawls, and the US bottom trawl and midwater trawl fisheries of the West Coast are required to have 100 percent observer coverage on-board.

Other countries, other rules. Some will have less observer coverage for large vessels, sometimes down to only a few percent of trips, and that reduces any effective enforcement accordingly. Really low-intensity observer programs are mostly for scientific purposes. We can estimate the catch of a protected species like sea turtles from observer data by dividing the number of turtles seen by observers by the percentage of fishing effort observed. However, the low sampling intensity for a relatively rare event such as a turtle incidental capture often means that the estimates are very uncertain and even biased. Because we lack any data to the contrary, for the purposes of this estimate, we must assume that an observed fishing boat behaves the same as an unobserved boat.

In general, on-board observers, especially when they are charged with enforcement or compliance-monitoring duties, are exposed to pressure from the crew, raising concerns about their safety or about the accuracy of the data they collected. For this reason, camera-based systems recording operations during fishing trips offer an alternative way of monitoring compliance.

Electronic Monitoring Systems, Automatic Identification System, and Port State Measures Agreement

Recently, Electronic Monitoring Systems, essentially cameras that record all fishing activities have become popular. They are much cheaper than on-board observers at least in developed countries, and can be used on smaller vessels that do not have room for observers. A set of tamper-proof cameras is connected to GPS systems, and when the vessel returns to port, the images are uploaded. The footage is usually put in storage, and only a sample is actually seen by the regulatory agency. In other cases, every fishing set is observed, and the information collected used for validation of catch data or estimation of bycatch. New research to automate the viewing and especially to try to get size and species composition by artificial intelligence may lead to total coverage of fishing vessels within the next 5 years.

The Automatic Identification System (AIS), originally designed for ship traffic management and safety, has been tracking 100,000 vessels of over 300 GT (roughly 100 feet or 33 meters in length) since 2002. By 2012, an estimated 250,000 ships were fitted with AIS that transmits identification and location online in real time. If you see a large ship in the harbor or at sea, your cell phone will tell you some basic identification information about the vessel. The original system relied on short-range VHF communications, but in recent years, satellites with low earth orbits have been able to also receive the AIS signals. With some limitations, AIS data are now used to map fishing effort of the larger vessels, as well as potential illegal fishing and transshipments. Of course, a captain can turn off the AIS transmissions or modify some of the information broadcasted (AIS is not tamper-proof), but that in itself will raise all sorts of suspicions among enforcing agencies.

Perhaps the most significant advance in enforcement has been the FAO Port State Measures Agreement that came into force in 2016. This agreement requires that foreign fishing vessels request permission to enter a port and report the nature of their fishing activities. It allows port officials to see if the vessel is on any black-lists of IUU vessels and to inspect the vessel and its paperwork. The Port State Measures Agreement has led to the arrest of a number of notorious IUU vessels.

Special Issues for the High Seas

Any enforcement outside the economic zones is extremely difficult. A persistent problem for everyone is unregistered vessels, which are, by definition, IUU. The home country within most RFMOs is in charge of enforcement in its own waters. By contrast, the flag States are responsible for the actions of its vessels at all times. If a flag State fails consistently with its obligations or refuses to cooperate with regional management arrangements, they might find it difficult to access main markets with their fishery products. When a ship comes into port, it must comply with the FAO Port State Measures Agreement that is designed to prevent known illegal fishing vessels from being able to enter ports, resupply, and ship their catch. There is a registry of illegal vessels and a tracking program to keep the list current that records the frequent name changes and reflagging to yet another country of registry. Nine RFMOs and Interpol participate in uncovering illegal fishing or transshipping of illegal fish.

Enforcement in Community-Based Management

Community-based fisheries with not too many boats fishing close to shore generally work best with social pressure. Anyone bending the rules with a little fishing in a closed area or on a closed day, or using a bit of dynamite instead of a net or hooks, will not be unobserved for long—and people talk.

Small-scale fisheries with lots of boats with many independent-minded skippers are possibly the most difficult to control. Obviously, one management strategy would be to limit how many boats can fish. Another one that is very effective when different groups of boats are restricted to specific areas is to paint them in area-specific colors. Any interloper is quickly spotted.

As I said at the outset of the chapter, compliance and enforcement are probably the most important elements of fisheries management. If the rules are not uniformly obeyed, there are incentives for all fishermen to bend the rules. If they know others are cheating, it will be harder to resist cheating too. Brian Mose, a very successful trawl skipper in British Columbia, told me that on-board observers have made his life better—he used to always try to find a way to game the system, but with on-board enforcement, he just sticks to the rules and has much less to worry about. But on-board observers are a luxury that only larger boats and richer fisheries can afford. I am convinced the way forward is continuous vessel location recording for all boats (including recreational boats) using modern technology, and on-board cameras for all commercial boats—at least in developed countries.

• •

FURTHER READING

An overview of IUU fishing. Baird R. 2004. Illegal, unreported and unregulated fishing: an analysis of the legal, economic and historical factors relevant to its development and persistence. Melb J Int'l L 5: 299.

A specific look at West Africa. Doumbouya A, Camara O T, Mamie J, Intchama J F, Jarra A, Ceesay S, Guèye A, et al. 2017. Assessing the effectiveness of monitoring control and surveillance of illegal fishing: The case of West Africa. Frontiers in Marine Science 4: 50.

A review of international law regarding IUU fishing. Jacobson J L. 1984. International fisheries law in the year 2010. La L Rev 45: 1161.

The story of unreported catch in the southern bluefin tuna fishery. Polacheck, T. 2012. Assessment of IUU fishing for southern bluefin tuna. Marine Policy 36: 1150–65.

CHAPTER 13

Seafood Certification and Non-governmental Organizations

U p to the 1990s, fisheries management in the United States and most industrial countries involved two groups: the government and fishermen. At a meeting with fishermen and government employees in 1982 that I attended, the keynote speaker was a very successful fisherman. His first sentence was, "The job of the fishing industry is to catch as many fish as possible." Looking at the government employees, he continued, "Your job is to keep us from catching too many."

This simple two-party relationship came to an end when a new factor entered fisheries management in the late 1990s. This profound change began in the United States and a few other wealthy countries and gradually spread to the rest of the world. Environmental non-governmental organizations (NGOs) increasingly secured a place for themselves as major negotiators in world fisheries management. Over the last 25 years, NGOs have owned and directed various forms of seafood certification and labeling and have become a powerful voice in the politics of fishing.

Northern Cod Collapse: The Seafood Certification and Marine Non-governmental Organizations (NGOs)

The closure of the cod fishery in Newfoundland on July 2, 1992, was the most shocking and consequential event in fisheries management since the declaration of the 200-mile limit in the late 1970s.

Cod had been fished off Newfoundland ever since boats from Europe were able to cross the Atlantic. Newfoundland is closer to the western edge of Europe than to most of Canada. It is so far east from even Eastern Canada, it has been given its own

Ocean Recovery: a sustainable future for global fisheries? Ray Hilborn and Ulrike Hilborn, Oxford University Press (2019). © Ray Hilborn and Ulrike Hilborn 2019. DOI: 10.1093/oso/9780198839767.001.0001

time zone—half an hour earlier. Basques from France and Spain and the Portuguese had fished there long before the English claimed it as a colony in 1583. Northern cod sustained a transatlantic fishery for a good 500 years. But by 1992, it had been driven to commercial extinction: 30,000 jobs in fishing and processing were lost, a calamity for a population of 580,000 that has, incidentally, declined ever since, and a wakeup call for fisheries managers around the world.

Nationally and internationally, fisheries managers began to strengthen their management systems, hoping to prevent similar collapses, and a series of scientific papers began to appear describing the decline of ocean fisheries. But perhaps more importantly, the cod collapse led to two major non-governmental initiatives that heavily influenced how the public perceives commercial fishing and in some cases how a fishery is managed. The direct results of the rise of NGOs are seafood certification and labeling, and efforts to influence fisheries management.

Certification and Labeling: What Is the Difference?

Both certification and labeling use consumer pressure to change how fisheries are managed and fishing industries behave. By identifying fisheries that meet certain standards, both certification and labeling advise retailers and consumers on what products they should sell or buy.

The Differences Lie in the Process of Reaching Recommendations

The best-known seafood certification system, the Marine Stewardship Council (MSC), certifies fisheries globally. The closest terrestrial food certification would be "certified organic." The MSC has set rigorous scientific standards of best practice for fisheries. A fishery must apply to be certified and go through a process of examinations and inspections of its practices and records that is arduous and protracted, can take years, and is costly. During that time, the MSC can make suggestions on changes or deny certification. Once a fishery is certified, it can display the MSC logo on its products.

Labeling is advice to consumers on the basis of an internal judgment by each advice giving entity. The terrestrial equivalent would be "locally sourced," or "from your family farm."

The *Seafood Watch* Program of the Monterey Bay Aquarium is the best-known advising entity in the United States and offers a web site, a smartphone app, and printed cards for your wallet that flag fish to avoid in red and note best choices in green and good alternatives in yellow.

The MSC is an independent nonprofit founded in 1996 by the World Wildlife Fund (WWF) and Unilever, a major fish retailer in the UK, to certify well-managed fisheries. The fishery must meet the MSC's management standards and show that its chain-of-custody system assures that the product on retail shelves is indeed fish from its fishery, because deceptive and downright false labeling of fish in retail and on restaurant menus is a persistent problem.

At the inception of MSC, many large retailers in the United States and Europe made commitments to sell only MSC certified seafood, which prompted many fisheries to submit to the certification process. As of December 2016, 296 fisheries were certified; 67 were under assessment; and 17 had their certification suspended.

Three principles are scored: (1) stock management, (2) ecosystem effects of the fishery, and (3) governance, policy, and adherence to legal management systems. There are seven scoring criteria for Principle 1; five for Principle 2; and nine for Principle 3. To be certified, a fishery must achieve an average score of 80 percent for each Principle and have no individual score below 60 percent.

Certification can be conditional pending changes during the first period of the process. It is believed that requiring changes during the process will improve the management of the fishery.

This evaluation is heavily weighted toward sustainability as a process rather than as a measure of ecosystem condition. It is also completely confined to the fish stock, ecosystem impacts, and the management system, without any consideration of social or economic impacts or environmental impacts beyond the local marine ecosystem.

The Monterey Bay Aquarium's Seafood Watch program is the best known of over 200 seafood guides that provide consumer advice. The program evolved from the *Fishing and Solutions* exhibit that ran from 1997 to 1999 and has four criteria: (1) impacts on the species under assessment, (2) impacts on other species, (3) management effectiveness, and (4) impacts on the habitat and ecosystem.

How Best Intentions Lead to Consumer Confusion

Neither MSC nor Seafood Watch includes social or economic impacts, and environmental impacts beyond the local marine ecosystem are equally missing. Central to Seafood Watch are criteria associated with environmental impacts of fishing and they are given much more weight than basic sustainability of the target species. This is perhaps best illustrated by the relative rankings of Western and Central Pacific Yellowfin tuna (as of February 8, 2018):

- "Best choice" if caught by pole-and-line fishing.
- "Good alternative" if caught by US longline or purse seine not associated with fish aggregating devices (FADs).
- "Avoid" if caught in association with FADs.

Pity the consumer searching for longline, purse seine, FAD, non-FAD information on a can of tuna.

The underlying reason for these categories is not stock size, nor management or sustainability of the target species, but only the bycatch of other species, and that is different for different fishing methods. Purse seining around FADs has significant bycatch of juvenile bigeye tuna and a range of other species. However, pole-and-line and longline fisheries produce three to five times more carbon per ton of tuna landed than do purse seiners. Consequently, if you were more worried about carbon than bycatch, purse seining would be green, and pole-and-line fishing red.

To compound a consumer's confusion, the underlying criteria of guides issued by individual NGOs are not at all transparent nor do they always make sense.

For example:

Alaska pollock from the Bering Sea is MSC certified. It is a "good alternative" but not the best choice from Seafood Watch, and it is red-listed by Greenpeace.

New Zealand hoki is MSC certified. It is rated as "do not eat" by New Zealand NGO Forest and Bird, whose guide is endorsed by WWF New Zealand. But— WWF Global says to look for the MSC label.

Different axes are being ground by different NGOs, keeping the conscientious consumer successfully in the dark as to their motivations.

Does Certification and Labeling Actually Matter?

It depends. You could argue that the sheer number of MSC certified fisheries and the changes made by conditions attached to certification have improved fisheries management. Seafood Watch's smartphone app is widely used, which can be seen as an indicator of its impact on fisheries.

You could also argue that the only thing that certification and labeling does is to point out which of the world's fisheries are already the best managed. Moreover, making a self-perceived informed choice is clearly a first-world, conscience-salving, rich-people's prerogative that creates a warm and fuzzy feeling about saving the planet. The not so rich eat imported, badly, or not-at-all managed fish from Asia.

Something Important to Remember

The world's hunger for protein is so great, whether you or I refuse to eat noncertified fish is globally utterly irrelevant. Fisheries that are MSC certified or otherwise labeled are only a small fraction of the total global catch. No fish's life will be saved by our informed choice since the global catch remains the same. The fish we choose not to eat will be sold to someone else.

Is It about Management or Is It about Access?

MSC certification certainly provides market access for some high-value fish products from well-managed fisheries, but I question how much it has changed management. Most conditions for MSC certification require changes in documentation rather than real changes in fishing methods of the already best-managed fisheries in the world. Some MSC certifications do require changes in the actual management. How catch and effort respond to abundance must follow a formal set of rules. MSC will not certify without such harvest control rules. Since the fisheries that met

MSC standards without having to change are already certified, it is the fisheries under evaluation now that may well require changes.

My initial skepticism stems from my experience with the certification of well-managed fisheries that were found wanting only in proper documentation. But I bow to my colleagues who have worked with fisheries that did indeed have to change management practices to get certified. Thus, it is probable that most future certifications will entail such MSC required changes.

Two Cases in Point

Alaska salmon are among the best-managed fisheries in the world and were one of the first fisheries certified by MSC. In 2009, Russian salmon were also certified by MSC, even though it was clear that the standard of management in Russia was much lower, with considerable illegal fishing among many other problems. But the certifying body found that the standards had been met. Alaskans were incensed and feeling that everyone knew that their management was better than the Russians' was, they withdrew from MSC. However, UK and German retailers made it clear that they would not buy Alaska salmon without MSC certification. Bowing to economic pressure, the Alaska salmon producers had to reapply, and were duly recertified by MSC. Did anything about Alaska salmon management change to get recertified? No—but the MSC label went back on their products.

Orange roughy is a very long-lived fish found in much of the temperate oceans. It was a latecomer to world fisheries, and yes, it really is a startlingly bright orange and can live for 100 years or more. It quickly became the darling of US restaurants because of its firm white flesh that could take any amount of a chef's creativity. As a cynic might put it, it is the perfect fish: colorless, odorless, tasteless, and with good flakes.

All that popularity was not beneficial. Stocks were rapidly depleted, and soon enough many of the fisheries had to be closed. Unsurprisingly, orange roughy became the icon of bad management with which NGOs could bludgeon fisheries managers for decades. New Zealand, the dominant producer of orange roughy, had taken the brunt of it when, owing to NGO pressure, the US market largely dried up. But there were plenty of others willing to buy roughy. It has become very popular in China, where consumers tend not to consult pocket guides before buying fish. Ironically, most New Zealand orange roughy catch is now MSC certified because the more recent scientific studies have shown that the stocks have recovered and now meet the standards of good management. Notwithstanding, roughy still languishes on some NGO red lists.

Fisheries Improvement Project Entry to the Marine Stewardship Council

The latest development in certification and labeling is the Fisheries Improvement Project (FIP),[1] intended to be a pathway to MSC certification and the much more important market access. Fisheries entering the formal FIP process can often get market access before reaching certification. Certainly, FIPs promise to help move many poorly managed fisheries up to MSC standards, but it may be too early to tell if they will have much impact on the fisheries of the world. As more and more fisheries seek to achieve the MSC standard, I would expect the FIP process to have a bigger role in changing management practice.

Follow the Money

In the aftermath of the cod collapse, vast sums of money became available from a number of foundations that started marine conservation programs. Let us see where the money comes from and where it ended up. The numbers we have were compiled by an evaluation of their programs that several major foundations commissioned.[2]

Where Did the Money Come From?

Between 2007 and 2009, many of the big donors came from Silicon Valley, where the Packard Foundation (Hewlett-Packard money) spent $79 million per year on marine conservation, and the Moore Foundation (Intel) $29 million per year. Other major foundations include The Walton Foundation (Walmart) $24 million per year and Pew (oil money) $27 million per year. Most of this money went to NGOs for their lobbying activities, but some was spent on science, and my colleagues at the School of Fisheries and Aquatic Sciences at the University of Washington and I have certainly profited from this interest in the oceans.

Environmental NGOs raised significant sums of money on their own. WWF spent $61 million per year from sources other than these foundations; Ocean Conservancy $13 million per year; Greenpeace $7 million; Wildlife Conservation Society $7 million; Oceana $6 million; and The Nature Conservancy $5 million. In total, the foundations from their endowments, and environmental NGOs from independent fund raising, spent a documented $317 million per year on marine issues. A number of major NGOs such as Conservation International were not included in the total.

Where Did the Money Go?

The numbers are from 2007 through 2009, and most was first spent in the United States and then in Europe, even though by 2007, stocks were recovering substantially in Europe and the United States. In Asia, Africa, and Latin America, where marine conservation was most needed, it received relatively little of the funding.

After the California Environmental Associates report pointed out the discrepancy between where the money was spent and where it was needed, these foundations have redirected their money from the United States to Asia and Latin America, and at least one, the Moore Foundation, has stopped funding most fisheries work.

What Did All Those Vast Sums Accomplish?

In the United States, much was spent on suing NOAA, the national fisheries management agency that had, at one point, one hundred lawsuits pending against it. I would argue that overall, these funds have had little impact on how fisheries are managed. In both the United States and Europe, fish stocks had already begun to recover long before, primarily because of legislative changes.

Certainly, there were some gains. Many of the lawsuits had been directed at the groundfish management in New England, and they undoubtedly hastened the shift to more conservative management policies (see Figure 13.1). Yet these are not major US fisheries, and they too had begun to recover back in the mid-1990s.

It had been the shock of the cod collapse that drove the rapid and influential changes made by governmental fisheries management agencies. The big foundation and NGO money concerned with fisheries did not materialize for nearly another decade.

A Closer Look

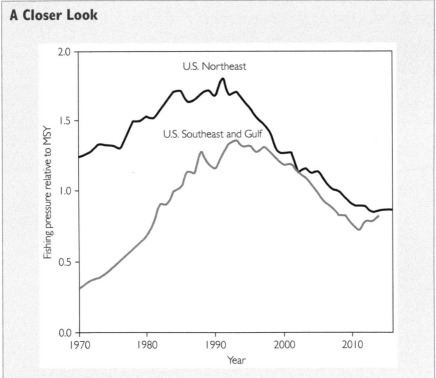

Figure 13.1 Trends in fishing pressure in the US Northeast and Southeast and Gulf since 1970.

The fishing pressure relative to the level that would produce maximum sustainable yield started declining in the mid-1990s because of changes in the US law. These reductions in fishing pressure began well before the well-financed NGO efforts.

A Miss and a Success

The Packard Foundation spent tens and perhaps hundreds of millions of dollars supporting research on Marine Protected Areas (MPAs), and lobbying for their implementation. Eventually, dozens of them were established in California. But after 15 years, there is no evidence that this set of MPAs has led to more fish or changed fisheries management in California. There are indeed more fish inside the reserves, but there are also fewer outside, and a recent study suggests the likely net result is no significant change.[3]

A major success for NGO efforts has been putting pressure on Japan, France, and Spain to improve management of Atlantic bluefin tuna. All three countries mostly eliminated illegal fishing and greatly reduced the take of bluefin. There are now signs of a major recovery in stock abundance that has allowed an increase in the legal catch.

Rapists and Pillagers of the Oceans—Why It Pays to Beat Up on Commercial Fishing

Ocean NGOs have been part of our lives for so long, we are vague about each one's purpose. All of them are for healthy fish populations and oceans, but when it comes to fishermen, sentiments range on a wide arc from love and cherish to hate and let them perish.

The Nature Conservancy and the Environmental Defense Fund tend to work with fishing groups as part of their strategy, whereas at the other extreme, Greenpeace opposes any industrial fishing whatsoever. Indeed, one of the founders of Greenpeace, Patrick Moore, who had a falling out and left, said that Greenpeace is no longer an environmental organization, but rather an anti-industrial one. He also criticized Greenpeace's use of scare tactics and disinformation, saying that the environmental movement "abandoned science and logic in favor of emotion and sensationalism."[4]

To understand the NGOs in the fisheries world, we must be aware that each one is trying to carve out its niche for recognition in a rather crowded group competing for donors. The major ocean NGOs have large staffs that need to be paid and retained, making fund raising perhaps their most important activity.

I once asked an acquaintance who worked for the US WWF why WWF New Zealand was quite so rabid about commercial fisheries in general and New Zealand hoki in particular. Despite MSC certification, WWF New Zealand is adamant about a "do not eat" dictum for hoki. He said that it all goes back to the sinking of the *Rainbow Warrior*, Greenpeace's flagship, in Auckland harbor by a husband and wife team of French secret agents in 1986 that killed a photographer. The *Rainbow Warrior's* mission had been to protest and document French nuclear testing in French Polynesia, particularly on the island of Moruroa, and the ship had been refueling in Auckland. Ever since then, the New Zealand public has had a particular affection for Greenpeace, and people line up for blocks to visit and donate to the cause whenever a Greenpeace ship docks in a New Zealand harbor. According to my acquaintance, it was really quite simple. If WWF New Zealand were to adopt any positions different from Greenpeace, New Zealand fundraising would dry up.

It really is all about the money, and to get it, each NGO needs to convince potential donors that there is a very serious problem and only they have the tools to solve it. Often short on science, some NGOs resort to the "emotion and sensationalism" that Patrick Moore found so objectionable.

One of the emotions used is despair. Some NGO web sites tell their story. NRDC: "Armadas of industrial fishing boats are wiping out our oceans' fish. The populations of tuna, swordfish, and other large species have fallen by 90 percent."[5] Greenpeace: "Nearly a third of commercial fisheries globally have already collapsed. We've already removed at least two-thirds of the large fish in the ocean, and one in three fish populations have collapsed since 1950."[6] WWF: "Overfishing and destructive, wasteful fishing practices are threatening the health of our oceans and food security for communities everywhere."[7]

Is It All Doom and Gloom?

No. Other NGOs offer science and hope. Oceana: "Save the oceans, feed the world."[8] Conservation International: "Transforming wild fisheries and fish farming. The ocean can provide food for billions – but only if we take care of it."[9] The Nature Conservancy: "Sustainable fisheries mean secure livelihoods, stable seafood supplies, strong coastal communities and a healthy ocean."[10] The Environmental Defense Fund: "From Namibia, Belize and Denmark to Australia, Chile and the United States, fishing rights systems are reversing overfishing, reviving coastal communities and bringing the oceans back to life."[11]

The Presence and Power of Ocean NGOs

NGOs are now active in almost every country in the world, hoping to change national legislation and management, and often working directly with small fishing communities. Each one brings its own range of approaches and toolkits.

The United States is undergoing a review of its national fisheries legislation, the Magnuson-Stevens Fisheries Conservation and Management Act, passed by the House in June 2018, but not yet by the Senate. The NGOs have a big role to play in this. One Washington insider told me, "the Democrats do what the environmental NGOs tell them needs to be done, the Republicans want to do whatever the environmental NGOs don't want." The Environmental Defense Fund, whose tool of choice is catch shares, such as individual transferable quotas and sector allocation, has been particularly successful at getting such methods implemented in many US fisheries. A range of NGOs has pushed successfully for declaration of large areas as no-take MPAs using the Monuments Act under the Obama administration. NGOs also have some representation on US fisheries management councils. Quite simply, NGOs are now among the most important actors in the fisheries management world.

Nowhere have antifishing NGOs had more success than in Australia. The US-based Pew Foundation has spent millions of dollars around the world promoting no-take MPAs, and Australia has proved to be fertile ground. As of 2017 over 30 percent of Australia's economic zone was in Marine Parks.

The political power of NGOs in Australia became apparent when a campaign launched by Greenpeace convinced the government to change national legislation to ban one specific fishing boat. That was the "super-trawler" controversy.

In 2012, an Australian fishing company found that it could no longer economically catch the 18,000 tons of its allocated jack mackerel quota in Tasmanian waters. Jack mackerel is a low-value fish primarily used for fishmeal and oil in aquaculture. 18,000 tons in commercial fishing is a relatively small amount, and, given the cost of fishing, the quota went unused.

In order to be able to use it, Seafish Australia chartered a Dutch fishing vessel, the *Abel Tasman*, to catch the jack mackerel. The *Abel Tasman* is a large factory ship that can catch and freeze the fish on board and sell the whole catch for human consumption in West Africa where there is a good market for it. But it was not to be. Sixteen NGOs, including Greenpeace and the Sea Shepherd Society, launched an intensive media campaign arguing that this super-trawler would destroy Australia's ocean. There was a full-page ad in Australia's major newspaper, a national day of action with rallies and flotillas of small boats, and an online petition that attracted 95,000 signatures.

Under Australian law, the *Abel Tasman* was perfectly legal to fish. But under this much pressure, the government buckled and introduced a bill that specifically banned this one ship. A stunned academic and fishery-science community voiced its concern about the triumph of emotion and sensationalism over basic science and good fisheries management policy:

> Whereas science is usually deployed in support of conservation in natural resource conflicts, in this case science based fisheries management advice took a back seat to vociferous protest by interest groups, perpetuated by the media (in particular social media).[12]

In New Zealand, the antifishing NGOs have also been particularly effective. In 2017, the prime minister announced the creation of the world's largest no-take MPA in the Kermadec Islands. NGO pressure in late 2017 caused the government to mandate hourly reporting of fishing vessel location via satellite tracking system and around the clock on-board camera coverage of all commercial fishing vessels. Surveys of the New Zealand public show that people earnestly believe their fish stocks are at serious risk of extinction, whereas the same people feel that on land, things are much better, despite more than 100 endemic species having gone extinct.

> In 2015, 96.8 percent of fish caught were from stocks that are not overfished. Public perceptions of fisheries and management are moving counter to scientific evidence as most fish stocks meet soft targets, and bycatch and numbers of trawls per year decrease.[13]

Both Australia and New Zealand are under serious threat that commercial fishing is losing the battle with the public over the "social license to operate." The public has been led to believe that commercial fishing cannot operate in a sustainable way with acceptable levels of environmental impact. A major motivation for seeking MSC certification in both countries is not primarily market access, but to bolster the commercial industry's arguments for their social license to operate. Environmental NGOs in both countries are not only adamantly against commercial fishing, but they have formed informal alliances with recreational fishing groups to whom commercial fishing is also anathema.

Antifishing NGOs have succeeded in making fishing regulations more conservative and have closed more and more areas to fishing. As a direct consequence, Australia is now importing 70 percent of its seafood. Fish consumption in both Australia and New Zealand has remained the same. But by setting such high standards for the environmental protection of their own fisheries where commercial fishing is concerned, and by importing such a high proportion of their fish, both countries are exporting the environmental impacts of fishing to importing Asian countries where management standards are low or nonexistent. Warm and fuzzy righteousness does have consequences.

The success of antifishing NGOs in Australia and New Zealand should serve as a warning to fishing industries in other countries where NGOs also have strong public following. The public, in general, needs to be made aware when and how industries are meeting high environmental standards through absolute transparency and documentation, high-profile advertising, and a strong social media presence with a focus on sustainability and food production. The kind of detailed reporting and observation of location and catch now coming into force in New Zealand is going to be the standard that fishing companies around the world will have to accept and implement.[14]

Even though major foundations shifted their funding to much of the developing world once they recognized that US and European fisheries are well-managed, there are still NGOs flogging the same old horse. As one industry wag put it, "they torture the best, and ignore the rest."

My Dust-Up with Greenpeace

I have what you might call a nuanced relationship with NGOs. Two of my closest collaborators and frequent coauthors are Ana Parma from Argentina, who serves on the board of directors of the Nature Conservancy, and Chris Costello, a professor at the University of California Santa Barbara, who holds a similar position with the Environmental Defense Fund. Both NGOs have supported my research in recent years and I work closely with many of their staff. So far so good. But...

I well remember the morning of May 11, 2016. When I turned on my computer to check email, I found a letter from Greenpeace that had been sent to the University of Washington and major journals, including *Science* and *Nature*, accusing me of hiding the fact that I had received funding from fishing industries. Greenpeace had begun a substantial publicity campaign in the United States, Europe, and New Zealand, to discredit my work, and for the first time, I made the front page of the

Seattle newspaper above the crease. Fishing industry money had indeed fully or partially paid for the research on thirty or so of my scientific publications, and that funding had been properly acknowledged each time. Before the dust settled, the University and the journals conducted thorough and time-consuming reviews and concluded I had complied with their policies regarding conflict of interest.

Exonerated, yes, but the stigma remains that fishing industry money is tainted, whereas government, foundation, and even NGO monies are considered clean.

A Sticky Subject: Research Funding

I do believe that the source of money does not matter when the work is done by academic scientists subject to peer review. Good scientists will do good work regardless of the funding. When colleagues in some NGOs were asked by reporters if they believed the Greenpeace accusations, the NGO staff supported my independence. Moreover, in almost all of the work I have done, I have had coauthors, such as Ana Parma and Chris Costello, who were not supported by industry funds.

The bitter truth is that without financial support, scientific research does not happen. An equally bitter truth is that there are specific interest groups who hire scientists to tell their story, and the history of scientists working for tobacco and oil does warrant concern about scientific independence.

I also believe that fishing industries have a responsibility to support fisheries research and that it is in their financial interest to do so. No one benefits more from sustainable management of fisheries than the fishing industry. Small-scale fishermen, whose employment and financial security depend on the fish they catch, have much more at stake in the sustainability of their fishery than any Greenpeace employee. I know several companies in large industrial fisheries of Alaska who are building new vessels that cost more than $50 million, an investment that depends entirely on sustaining the fisheries of Alaska. In today's world, if their fishery collapses, they are ruined because there is no other fishery to move to.

Disaster Is Good Business

Fisheries disasters and bad news are good for two groups of people: academics like me and environmental NGOs who raise funds from the public to "save the oceans." The more bad news there is, the more research money becomes available from governments and foundations. The biggest boost in research funding for our department was due to the Exxon Valdez Oil spill in 1989. Both Exxon and state and federal agencies immediately released millions of dollars. I was recruited by the State of Alaska as part of a team to evaluate the impacts of the spill. Money was no object since Exxon would have to reimburse the state for every penny.

If the Canadian cod and other fisheries had not collapsed, and there had been no overfishing in the 1990s, there would have been no cause for foundations to start marine conservation programs nor for NGOs to raise money from the public. Knowing this, it is tempting for some NGOs to ignore the scientific data that fish stocks are increasing and generally healthy in the United States and Atlantic Europe, and continue their pleas for money by crying alarm about the rape and pillage of oceans by commercial fishing.

Similarly, if the public in the United States, Europe, New Zealand, and Australia stopped believing that their fish stocks are collapsing, donations to doom-and-gloom NGOs would disappear.

Of course, there is no shortage of fisheries problems in the world, and most major US foundations have switched course to concentrate on improving management in developing countries. The challenge will be to persuade donors to fund fisheries issues thousands of miles away.

Finally, there is a very worrying trend that has been described as naked extortion, when one NGO is paid to provide advice to retailers on what to sell, and if they do not comply, a different NGO is called on to picket their stores. This has been thoroughly described by Wilson[15] as happening in the wood products industry. He argues that NGOs are now pushing to be the ones who determine how a resource-based industry is managed. This is certainly happening more and more in fisheries too. Legislators are influenced on how to write laws; presidents and prime ministers manipulated into a race on who can declare the largest piece of ocean as a no-fishing zone. One could certainly never say that the fishing industries have not also lobbied legislators, but their success has been minimal.

My concern is that the interests of both NGOs and the fishing industry are not necessarily in the public's interest. Therefore, the public needs to be equally skeptical of attempts by either group to determine how fisheries are managed.

FURTHER READING

A study of tuna FAD fishing. Ardill D, Itano D, and Gillett R. 2011. A review of bycatch and discard issues in Indian Ocean tuna fisheries. Indian Ocean Commission: Smartfish working papers.

The report on foundation spending in marine conservation. California Environmental Associates. 2012. Charting a course to sustainable fisheries. https://www.ceaconsulting.com/casestudies/charting-a-course-to-sustainable-fisheries/.

New Zealand public perception of fisheries and other conservation issues. Hughey K F, Kerr G. N, and Cullen R. 2016. Public perceptions of New Zealand's environment: 2016, EOS Ecology. Lincoln, New Zealand: Lincoln University Press.

Patrick Moore's critique of Greenpeace. Moore P A. 2010. Confessions of a Greenpeace dropout: The making of a sensible environmentalist. Vancouver: Beatty Street Publishing.

The thesis that evaluated the impact of MPAs in California. Ovando D. 2018. Of fish and men: Using human behavior to improve marine resource management. PhD Thesis Bren School of Environmental Science and Management. University of California Santa Barbara, Santa Barbara, California.

A comparison of seafood guides. Roheim C A. 2009. An evaluation of sustainable seafood guides: Implications for environmental groups and the seafood industry. Marine Resource Economics 24: 301–10.

Scientific comment on the super trawler issue. Tracey S, Buxton C, Gardner C, Green B, Hartmann K, Haward M, Jabour J, et al. 2013. Super trawler scuppered in Australian fisheries management reform. Fisheries 38: 345–50.

A critique of NGO efforts to control resource management. Wilson T. 2012. Naked extortion? Environmental NGOs imposing [in] voluntary regulations on consumers and business. Journal of Oil Palm, Environment and Health 3: 16–29.

CHAPTER 14

Ecosystem-Based Management and Marine Protected Areas

The Search to Define Ecosystem-Based Management (EBM)

Ecosystem-based management (EBM), the present darling of the fisheries manage-ment world, is elusive. Much like good art, while no one can really define it, most of us believe we will know it when we see it.

NOAA does have a definition. "The goal of ecosystem-based management is to maintain ecosystems in a healthy, productive, and resilient condition so they can provide the services humans want and need." This sounds a lot like *sustainability* (see Chapter 2)—managing the oceans so that benefits to humans do not decline over time. Fisheries managers certainly thought they had been trying to do that for decades. So what is new?

The Persistence of Old News and the Inconvenience of Fundamental Questions

There is steady standard finger pointing at fisheries management that becomes almost generic. "Many of the world's fish populations are overexploited, and the ecosystems that sustain them are degraded."[1]

True, overfishing had been a global problem in the 1990s, and it remains so in some areas of the world. But can we not finally acknowledge real progress, that overfishing has dramatically declined wherever fisheries are intensively managed, like the United States and much of the world? Twenty years have gone by—time to move on, wouldn't you think? Apparently not, because it is precisely in the United States where the cry for EBM is the loudest.

Ocean Recovery: a sustainable future for global fisheries? Ray Hilborn and Ulrike Hilborn, Oxford University Press (2019). © Ray Hilborn and Ulrike Hilborn 2019. DOI: 10.1093/oso/9780198839767.001.0001

The need for it comes from our past focus on single-species management that in the narrowest sense looks at adjusting the harvest of a target species to maximize the long-term yield of that one species. In theory, no consideration is given to impacts on other parts of the ecosystem. And the criticism is still all about the old failures of single-species management. I strongly suspect that if we had not made such a muddle of that, there would be much less call for EBM.

The world is crowded with creatures—of course, fishing affects more than the target species. Fishing gear often catches birds, mammals, turtles, and nontarget species of fish, which we discussed in the chapter on bycatch and discarding. Bottom contact gear like trawls, dredges, and longlines alters the sea floor. When we harvest one particular species, we reduce its abundance and thereby ease up on its prey, which, in turn, is having a fine time gorging on whatever it eats, but we take food out of the mouths of predators who feed on our target species. As always, one thing leads to another.

But it is not only the creatures—how we manage fisheries also affects human communities. Do we want fisheries to maintain small, isolated communities that depend on fishing or do we want fisheries to provide maximum wealth for our country? That is something we, the people, must decide and codify through our legislature.

EBM Has as Many Facets as a Well-Cut Diamond

Nearly all countries that manage fisheries intensively have recognized that we must move beyond single-species management to something more comprehensive, and that has been an integral part of traditional fisheries management systems in various places. Beginnings have been made by some agencies that have reduced catch through various means in order to protect sensitive habitats.

Imagine if you will, three blind men exploring an elephant. One runs his hands along a cool, smooth, and slightly curved surface in the round; another seeks to make sense of tough and deeply wrinkled skin with the odd coarse hair sticking out; whereas the third one is more explored than exploring by a curiously movable tube with a soft, short, and restless finger. That is pretty much how EBMs are seen— lots of detail without the whole picture.

We could, for instance, declare our EBM to be reasonably simple by addressing concerns about bycatch, forage species, and habitat modification as part of traditional single-species management. Or we could center our EBM around trophic connectivity, with the ultimate goal of accounting for species interactions using ecosystem models rather than single-species models. Then again, we could take a comprehensive view that encompasses the broad impacts of society, such as land use, national economic policy, and human population growth, when managing marine ecosystems.

Slight hitch here. EBM will be at least as demanding, and more costly, than single-species management. If fisheries agencies have been unsuccessful at implementing single-species management, should we expect them to do better with a necessarily more complex and expensive system?

I distinguish between core and extended aspects of EBM.[2] The core has three primary features:

1. Doing single-species management right, i.e., keeping fishing mortality at or below the level that will produce maximum sustainable yield and keeping fleet capacity in line with the potential of the resources.
2. Preventing bycatch of nontarget species through gear modification and providing incentives for bycatch avoidance or through area and seasonal closures (see Chapter 9).
3. Avoiding habitat-modifying fishing practices primarily by closing areas or banning specific fishing methods or gears (see Chapter 10).

Extended EBM is characterized by a consideration of trophic interactions and area-based management. Even though the three core elements are already widely used, we should move beyond them to broader ecosystem impacts of fishing. Let us go out on a limb and include people.

But, before We Get to People, What about the Broader Impacts of Fishing?

Some extended EBM is easy. Good, appropriate use of the tools of single-species management already reduces ecosystem impacts considerably. Rather than aiming for total maximum sustainable yield, we can get close to it with quite a few different harvest rates—a range that I call pretty good yield (PGY). Any one of a number of harvest strategies that will fish down a stock to between 20 to 50 percent of its unfished size will get you that PGY. If you keep to the lower end of that range of harvest rates, the biomass will stay higher and any ecosystem-wide impacts will be much lower.

Obviously, fishing affects ecosystems, and there are, at present, no guiding principles, internationally or in national jurisdictions, on the appropriate trade-off between sustainable yield and ecosystem impact. In single-species management, MSY has become a known reference point, whereas in EBM there is no agreement on where that sweet spot might be.

In fact there is no real consensus on what actions can be considered EBM. In a recent paper,[3] a survey was conducted of a wide range of people involved in fisheries management, and it asked if certain specific actions were part of EBM. The answers were all over the place. When asked about potentially removing sea otters from specific sections of the California coast to protect sea urchin and abalone fisheries, there was strong agreement and strong disagreement. Answers to whether it should be part of EBM to control predators to increase the abundance of target species were equally divided. However, almost all respondents agreed that both closing areas to protect sensitive habitats and reducing bycatch were part of EBM.

Almost all fisheries management agencies and community-based management around the world employ some form of area-based management such as closed areas to protect spawning stocks, juvenile fish, or sensitive habitats. We will talk about MPAs later in this chapter. Area-based management is already part of the mix and will certainly continue to expand.

On the whole, EBM is seen as less harsh on marine ecosystems than single-species management, but it all depends on how it is implemented. It is quite possible that in order to get the best production of goods and services from marine ecosystems, ecosystem knowledge and models will be used to manipulate the system to that end.

Obviously, we can deliberately overexploit a low-value species that preys on or competes with a more desirable high-value species. Terrestrial equivalents would be shooting lions, wolves, or bears and plowing up native habitat for a much higher return in food production. In marine ecosystems this happens already with shellfish culture, especially so in Asia. Will this kind of deliberate transformation inevitably become an ever-increasing part of area-based ecosystem management?

Marine Mammals—That Is Where the Sparks Will Fly

Somehow, harvesting marine mammals sounds a lot more ominous than harvesting fish. Death by gun shot is more real than death by net. And yet, getting rid of marine mammals that compete with fisheries is another facet in the crystal of EBM. Populations of sea lions and seals have grown substantially and have definitely had an unwelcome impact on many fish populations, even on endangered killer whales in Puget Sound. There is already a limited harvest of sea lions that are eating altogether too many endangered salmon in the United States, and in Eastern Canada, cod-eating seals are in the firing line. How killing or removing marine mammals to protect fisheries will play out in the conflict with the US Marine Mammal Protection Act will be an interesting process to follow. Worldwide, we may well see large-scale harvesting of some marine mammals in the name of EBM.

What about Us, the People—Where Do We Come in?

It seems to be implicit in most scientific thinking that EBM is all about the natural ecosystems, and that regulations of harvest and fishing gear are the key control variables. Yet, allocation of access to fishing is definitely another tool of fisheries management. That is where the "human element" disrupts the natural ecosystem. It is fair to say that acrimony and dispute over who gets to fish and where eats up as much or more management energy as does comparatively easy harvest regulation.

Lately, there is more and more evidence of an intertwining between how fish are allocated and its ecosystem consequences. Questions will arise of what value we assign to artisanal and small-scale coastal fisheries as opposed to efficient and profitable industrial fisheries. As a society, are we willing to forgo some of maximum food production and maximum profit to keep small-scale fishing alive? Should we require EBM to routinely encompass the impacts of fisheries management on human and aquatic communities or is EBM going to be confined to aquatic ecology?

That Again Is Something That We, the People, Must Decide

In summary, many fisheries jurisdictions are deeply engaged in EBM, and we can move toward improving ecosystems without losing much yield. However, we will have great difficulty in applying EBM beyond its core components without firmer policy guidance on the social objectives of fisheries and their impact on both marine ecosystems and human communities.

Ecosystem Protection and Marine Protected Areas (MPAs)

Another tool of fisheries management, often equated with EBM, is the establishment of marine protected areas (MPAs).

The very definition of *MPA* is slippery. How much protection and from what? Fundamentally, *protection* ought to mean just that—an area of the sea floor protected from mining or oil exploration is already technically an MPA. What concerns us here are no-take MPAs and no-entry MPAs. No-entry is the strictest form of protection: nobody crosses its boundaries. It is found in some places on Australia's Great Barrier Reef. No-take protection, of great importance to fisheries means do not take anything from within the boundaries. In short, do not fish there.

No-take MPAs are, for the most part, coastal, either around islands or along mainland shores. They protect its marine inhabitants from fishing—but not from ocean acidification, global warming, agricultural run-off, oil spills, silt, debris, plastic junk that arrives with rivers and storm water, pleasure boats, snorkelers, swimmers, surfers, or divers. Just from fishing.

So Where Do the Fishing Boats Go? They Go Wherever the Fish Are Outside the MPA

MPAs are a trend—countries are in competition on who can legislate the biggest and the most MPAs. There are bragging rights involved, and fund raising too.

In September 2016, around 5,000 participants, including government leaders, heads and representatives of non-governmental organizations (NGOs), and conservation scientists met at the IUCN World Conservation Congress to discuss how to reconcile preservation of the world's natural heritage with human development. Twenty-three NGOs, including the Pew Charitable Trusts, the Natural Resources Defense Council, and the Cousteau Society, successfully tabled a motion that called for achieving a target of 30 percent of coastal and marine areas to be fully protected by 2030.

Wow, That Is a Lot of Ocean—What Would Happen?

Obviously, there would be more fish inside than outside. In areas that are closed to fishing and where there is effective enforcement, the density of fishes within the reserve goes up by 166 percent. Impressive, no? However, we do not really know

A Closer Look

Marine Protected Area (MPA) models have shown repeatedly that regional abundance will increase if stocks have been greatly overexploited but will not increase when stocks are not overexploited. The top panel of Figure 14.1 shows the trend in abundance of fish (total inside and outside of reserve) when the stock is fished much too hard and is headed toward total collapse. In year 50, 30 percent of the area is put in reserves, and the total population (thick black line) starts to increase. Abundance inside the reserve (thick gray line) rises rapidly, but abundance outside the reserve declines to near zero as all of the fishing effort is moved and the harvest that was formerly taken from all areas is now taken from 70 percent of the area.

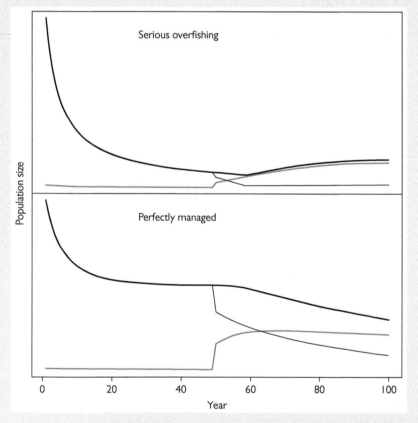

Figure 14.1 The trend in population abundance when no-take marine protected areas are put in place if the population is overfished (top panel) and if the population is well-managed (bottom panel). Data from Hilborn et al. 2006.

The lower panel shows what would happen if the stock was perfectly managed to MSY before the reserve. When the reserve is put in place, abundance inside the reserve (solid gray line) rises, but abundance outside the reserve declines because of excess fishing pressure outside. The net result is a decline in abundance.

A Closer Look

The failure of MPAs to increase overall abundance has recently been shown to be the case in what is perhaps the best-studied MPA network, set up in 20003 in the Channel Islands of California.

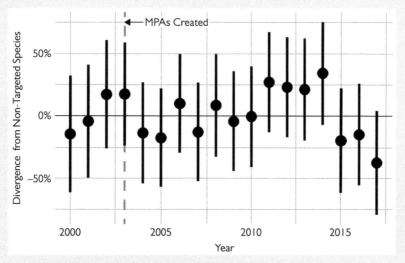

Figure 14.2 Estimated trend in abundance of total targeted fish species in and around the Channel Islands of California. Vertical line shows creation of marine protected areas.

The 2018 PhD thesis by Daniel Ovando, University of California Santa Barbara, estimated the total change in the abundance of targeted fish species inside and outside the Channel Island reserves. Ovando used the trend in nontargeted species as a control for the environmental changes. Figure 14.2 from his thesis shows the estimated relative abundance of targeted species compared to nontargeted.

The vertical bars show the confidence limits. What he found is that there has been no significant increase in abundance of targeted species, but that in recent years, the best estimate is that abundance has actually decreased.

whether MPAs where fishing is either tightly restricted or banned outright enhance or damage biodiversity on the outside. The fishing boats do not stop fishing; they take their fishing pressure elsewhere. Consequently, we expect to see biodiversity benefits inside the reserve, but may see negative impacts outside from increasing fishing effort.

Calls for MPAs began in earnest during the 1990s, when overfishing was common in most of the developed world and collapses of fish stocks repeatedly made headlines. In the early literature, marine ecologists often assumed that biodiversity could flourish only inside protected areas. In 2002, for example, it was suggested that 40 percent of the oceans should be set aside for reserves, based on the assumption that there would be no fish outside to reproduce.[4]

Most ecologists and conservationists now accept—in theory at least—that, even if MPAs were to cover 20 percent of a region, wherever there is effective fisheries management most of the ocean's biodiversity will exist on the outside. Yet when it comes to actually setting policy, the use of MPAs as a sort of security blanket increasingly dominates conservation strategy.

Why I Have Been a Consistent Critic of MPAs as the Dominant Conservation Strategy

The biodiversity of a whole region is better protected when 100 percent of the area is governed by good fisheries management than by closing 10 percent, 20 percent, or 30 percent to fishing inside MPAs. It is not clear whether MPA advocates willfully ignore or truly do not understand that the abundance of fish in a region will not go up when MPAs are put in place if the region's fisheries were regulated to maximize sustainable yield. Of course, there will be higher abundance inside the MPAs, but all the former inside boats will join the usual outside boats, and together they will drive down outside abundance.

MPAs have become the darlings of quite a few environmental NGOs and conservation funding groups. Together, the Pew Ocean Programme, the World Wildlife Fund, and Greenpeace, for instance, have spent millions of dollars in the past 10 years promoting MPAs all around the world.

With an unassailable belief in and focus on MPAs, starting in the 1990s, its proponents did not notice or would not acknowledge that national and international fisheries agencies were already developing the tools and legal frameworks to reduce overfishing, rebuild fish stocks, and protect marine biodiversity.

Thanks to fisheries management, overfishing has largely stopped in the United States, the EU Atlantic fisheries, New Zealand, Australia, Iceland, Norway, and Canada, and where stocks had declined below target levels they are now rebuilding (see Chapter 5). Over the past decade, major Latin American countries, including Peru, Argentina, and Chile, have started implementing strategies that have reduced the proportion of stocks fished too hard from 75 percent in 2000 to 45 percent in 2011.

Countries with effective fisheries management now make up 45 percent of global catch, and their fish stocks are stable or more often increasing. They do not need MPAs to protect fish stocks. Abundance of fish would not go up. Yet, it is precisely there where a major push for ever more MPAs is under way. Let go of it. Redirect effort and money to the fisheries in the Mediterranean, South and Southeast Asia, and Africa, where the marine environment is not protected through fisheries management nor are there significant no-take areas. I would be the first to cheer on efforts to get MPAs established there.

So, Can We Relax Now Where We Have Good Fisheries Management?

No, not on your life. Look around you, neither MPAs nor fisheries management alone can protect marine biodiversity from global threats like ocean acidification and climate change, nor can it avoid the damage from sources that fisheries management cannot regulate like land-based run-off, oil spills, plastics, particulates from air pollution, ship traffic, and the overwhelming amounts of just plain garbage.

When Will MPAs Increase Catch and Abundance?

The evidence we have so far is that only when fishing pressure is excessive and the shape and size of reserves is tuned to the movement of many species can MPAs increase fish abundance throughout a region and potentially increase or maintain food security. But because different species do not move in synchrony, it is unlikely that we can have it all. No MPA will ever benefit a lot of species inside and outside its boundaries.

Since fish go through various life stages and those stages do not stay put the size of the reserves, in order to benefit catch, must ideally be tuned to the adult and larval movement of the resident species. If an MPA is too small, fish will move in and out of the reserve frequently enough to be caught outside. If it is too big, there will not be a spillover of eggs, larvae, or adults to the outside. Although garnering no accolades from the public, the less expensive fisheries management, unrestrained by reserve boundaries, can precisely target regulations to benefit individual species.

On the Efficiency of Tools

We have the management tools, and they are being used successfully. If a stock is overfished, reducing fishing pressure is the answer rather than shifting it around to the outside of MPA boundaries. Where the threat is bycatch of non-target species, whether sharks, birds, mammals, or turtles, technical solutions and gear-specific time and area closures have been effective. Where damage to sensitive habitats is the issue, those areas can be closed specifically to the damaging gear.

Precision Targeting through Management

Fisheries management regulations generally recognize the spatial heterogeneity of fish populations and habitats and have the means to be highly selective as to which areas to open to what kinds of fishing at what specific times of year. These are all standard practice in good fisheries management, and where good management happens, there is no need for no-take MPAs. Fisheries agencies can better protect biodiversity and associated benefits by implementing and enforcing science-based regulations than simply designating no-take areas.

A reminder: technically, any managed area is considered an MPA. Only recently has the literature and public perception shifted to view only permanent no-take areas as providing effective protection.

A Closer Look

At international political discussions of MPA targets, there is stark disagreement about whether the target should be 10 percent, 20 percent, or 30 percent, and more importantly, about what counts as an MPA. 168 countries have signed the Convention on Biological Diversity, which has specific targets for MPAs under Target 11.

> Target 11: By 2020, at least 17 per cent of terrestrial and inland water areas and 10 per cent of coastal and marine areas, especially areas of particular importance for biodiversity and ecosystem services, are conserved through effectively and equitably managed, ecologically representative and well-connected systems of protected areas *and other effective area-based conservation measures*, and integrated into the wider landscape and seascape.

MPA proponents argue that only a permanent no-take MPA counts toward the 10 percent target for marine areas without acknowledging the "other effective area-based conservation measures." I, and most scientists working in fisheries management, would argue that closing areas to fishing gears that impact vulnerable marine ecosystems certainly constitutes an effective area-based conservation measure. Similarly, time-area closures to prevent bycatch and closing of spawning or juvenile rearing areas are all effective area-based conservation measures.

What about Alaska—Is It In or Out?

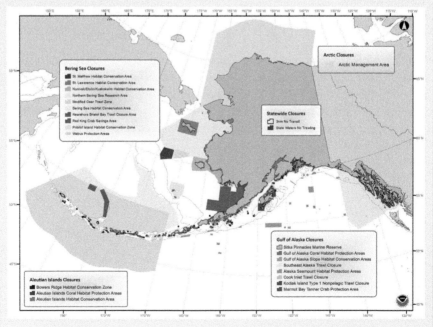

Figure 14.3 A map of Alaskan ocean waters showing the many kinds of protection that are in place.

The fisheries management system in Alaska uses intense area-based management (Figure 14.3). All of southeast Alaska and many other areas are closed to bottom trawling. Many gears are excluded from many other areas, as shown in the shaded areas of this graph. All of the Arctic Ocean is closed to fishing at present—awaiting scientific evaluation of the potential to have well-regulated fisheries. But there are almost no permanent no-take MPA, and thus the MPA fundamentalists feel justified to argue that Alaska has no areas that meet Target 11.

So in conclusion, I do see a valuable role for MPAs as tourist attractions and as reference sites for sedentary species that do not wander off. When other forms of fisheries management are not possible but MPAs can be implemented and enforced, they can benefit biodiversity and fisheries in cases of serious overfishing. Wherever there is no effective fisheries management but where MPAs could truly be enforced, networks of MPAs would be a positive step. Where there is high uncertainty, MPAs can act as a buffer against that uncertainty and variability.

Fish Do Not Belch Methane

However, always remember that MPAs only protect the ocean from legal, regulated fishing. Even if that elusive 30 percent target were achieved, it would not provide protection from the real threats, nor would it bring particularly good protection from fishing. The protein lost to us when large marine areas are closed must inevitably be replaced with protein from the land, with all the associated high environmental costs. Both biodiversity and food security will generally be better served by expanding fisheries management, not by establishing more no-take areas.

• •

FURTHER READING

A paper arguing for the importance of forage fish to ecosystems. Alder J, Campbell B, Karpouzi V, Kaschner K, and Pauly D. 2008. Forage fish: From ecosystems to markets. Annual Review of Environment and Resources 33: 153–66.

A classic paper on EBM. Francis R C, Hixon M A, Clarke E, Murawski S A, and Ralston S. 2007. Ten commandments for ecosystem-based fisheries scientists. Fisheries 32: 217–33.

The FAO description of ecosystem approach to fisheries. Garcia S M, Zerbi A, Aliaume C, Do Chi T, and Lasserre G. 2003. The ecosystem approach to fisheries. FAO Technical Paper no. 443: 71.

A conservationist's view of what works in MPAs. Giakoumi S, McGowan J, Mills M, Beger M, Bustamante R, Charles A, Christie P, et al. 2018. Revisiting "success" and "failure" of marine protected areas: A conservation scientist perspective. Frontiers in Marine Science 5: 223.

My own take on how to protect marine biodiversity. Hilborn R. 2016. Marine biodiversity needs more than protection. Nature 535: 224–6.

My take on when MPAs are effective. Hilborn R. 2017. Are MPAs effective? ICES Journal of Marine Science 75: 1160–2.

The role of protected areas in precolonial fisheries management. Johannes R E. 1982. Traditional conservation methods and protected marine areas in Oceania. Ambio 11: 258–61.

A major synthesis paper on changes of fisheries inside MPAs. Lester S E, Halpern B S, Grorud-Colvert K, Lubchenco J, Ruttenberg B I, Gaines S D, Airame S, et al. 2009. Biological effects within no-take marine reserves: A global synthesis. Marine Ecology Progress Series 384: 33–46.

Another overview paper on EBM. Link J. 2002. What does ecosystem-based fisheries management mean? Fisheries 27: 18–21.

A call for the importance of EBM. Pikitch E K, Santora C, Babcock E A, Bakun A, Bonfil R, Conover D O, Dayton P, et al. 2004. Ecosystem-based fishery management. Science 305: 346–7.

CHAPTER 15

Enhancement and Aquaculture

Humanity seems to have an overwhelming desire to produce more fish—it is almost magical that a handful of eggs can turn into a tankful of fish. Both ancient Chinese and Aboriginal Australians had perfected the rearing of fish, capturing small fish and feeding them until they are ready to eat. The real technological breakthrough came much later, when hatchery technology succeeded in mixing eggs and sperm to produce millions of baby fish, and with it came the almost unlimited potential to produce ever more fish.

Hatcheries: The Promise of Abundance Forever and Ever

As fish stocks decline and the political, social, and economic pains of ever-reducing harvests increase, the idea of hatcheries becomes irresistible. Why regulate fisheries when you can produce more fish in a hatchery? Loren Donaldson, a professor in my department long before I arrived, and who loved hatcheries, once said, "The philosophy of scarcity never fit my concept of production. We have a natural resource that can be used to the fullest. I'd like to make so many salmon there wouldn't be any need for regulation."

Making more fish is by now quite common. We can make more fish in hatcheries and release them into the wild where there are few or none at all and cross our fingers that some survive to edible size. Or we can eliminate the uncertainties of the wild and rear fish under controlled conditions from eggs to table—that is called aquaculture, the fastest growing form of food production that already exceeds both global beef produced and wild fish caught. We will talk about that later.

When we release hatchery-reared fish into the wild, it is called stock enhancement or ranching. Such releases may be designed to provide trout for anglers in reservoirs, lakes, and streams that are well beyond the trout's natural range. Or the

Ocean Recovery: a sustainable future for global fisheries? Ray Hilborn and Ulrike Hilborn, Oxford University Press (2019). © Ray Hilborn and Ulrike Hilborn 2019. DOI: 10.1093/oso/9780198839767.001.0001

aim may be to help rebuild stocks that were severely depleted—such as the Redfish Lake sockeye salmon that were down to the last handful. Those fish were brought to a hatchery, their progeny was reared for several generations, and eventually thousands were later released back to where they came from. There is also marine stock enhancement. Hatchery produced fish are released into the wild to supplement wild production. These fish are highly relevant to the management of marine fisheries, and in some places, they are a major part of the fisheries management strategy.

Enhancement: History

Hatchery technology, producing viable juveniles from eggs and sperm, called milt when it comes from fish, and rearing them to a size where they can be released into the wild, or eaten, was perfected in the eighteenth and nineteenth centuries. Mix eggs with milt, provide suitable water and temperature, and baby fish will emerge. It worked for trout, salmon, cod, lobster, and quite a few other species. Serious experimental programs started in Europe for cod and in the United States for trout and salmon.

It was amateurs of the Age of Enlightenment who first reared fish in hatchery conditions. In 1747, Ludwig Jacobi, a Westphalian army officer, spawned and reared trout. Many others followed in both Europe and the United States. The first institutional hatcheries were built in France in 1850, to bring back fish to the Rhine and Rhone Rivers. The first federal fish hatchery in the United States was opened in 1872 in Northern California and was designed to ship fertilized salmon eggs around the country and help establish new salmon populations. Not far from this hatchery, rainbow trout were discovered, a brand new species that could be reared easily: so easily that the thousands of rainbow trout populations all around the world have their origin at that very hatchery.

The early marine fish hatcheries did not attempt to feed the newly hatched fish, but simply put them out into the wild as eggs or larvae, a practice that continues. Even though evaluations in Europe and North America failed to show that any of these fish survived, efforts to shut down the hatcheries came very slowly. In Canada, salmon hatcheries were not closed until the 1930s. In Norway, the cod hatcheries hung on until the 1970s. The appeal and promise of this simple technology was just too great—we know how to make fish—here are millions of them, let us continue. And continue they did.

Eventually, good science and better technology came to the rescue. The first critical step was nutrition and control of disease. The baby fish need to be reared until they are big enough to have a chance to survive in the wild, and that requires finding the right feed and keeping them free of diseases. Survival after hatching is the biggest hurdle, and most fish are very specific about what they need to survive. In the 1960s and 1970s, hatchery technology was refined, and there are now well-established protocols for rearing dozens of marine species to almost any desired size.

The next key scientific advances were in ecology and economics. Life history and ecology must be understood enough to determine when and where to release the juvenile fish. On the one hand, the larger the fish the better their chance of survival, but on the other hand, the longer it takes to release the juveniles, the more it costs.

The ecological theory of enhancement is quite simple—marine fish produce many thousands or millions of eggs, but very few survive the first weeks or years of life. If you can rear fish in a hatchery past that perilous first period of life, lots of fish can be caught later. The second ecological factor is carrying capacity, how many of any organism a specific environment can support. For enhancement to work, the environment into which fish are released must be able to support them until they are big enough to be legally caught. In the case of the rainbow trout releases into fish-free lakes, as long as there were enough invertebrates that trout like to eat, those lakes were chock full of unused carrying capacity. By now, not quite 10 percent of freshwater fisheries depend heavily on hatchery fish to augment their stocks.

As to the ocean, if we assume that a species may have been overfished, we naturally also assume that there is unused carrying capacity to be filled with our hatchery fish that can now be found all around the world. At last count, perhaps one hundred species are used for some form of hatchery enhancement and they have become important in the marine fisheries of Japan and China, as well as in the Pacific salmon fisheries.

Hatcheries for Pacific Salmon

Pacific salmon hatcheries in the United States and Canada began in the late-nineteenth century but were confined to releasing fertilized eggs or newly hatched larvae. The main motivation was the decline in the catch of some major rivers, and the increasing loss of freshwater habitat due to cities, farming, logging, and dam construction. Jay Taylor is a historian who has documented the history of salmon hatcheries in the nineteenth century,[1] and he describes the mix of technological hubris and political desire to let habitat loss and overfishing continue as the key ingredients in hatchery efforts in that era. We now recognize that these efforts were totally ineffectual, and few of any of those fish survived, and almost all of the early hatcheries were closed.

But the technical innovations in rearing technique and feed changed everything. By the 1960s, salmon could be reared to any size, and some of those released into the ocean did survive. Various marking techniques were developed so that hatchery-reared fish could be identified when caught. In British Columbia, modern hatchery production began in the 1960s, and by 1990, about 700 million juveniles salmon were released.[2] This declined to half by 2000.

I take some credit for this decline. Back in 1975, I was hired by what was then the Canadian Department of Fisheries and Environment and mostly worked on evaluating the hatchery program. In 1980, I began working at the University of British Columbia, funded by a grant by what was then called the Salmonid Enhancement Program. The rapid expansion of hatchery production in the 1960s and 1970s was fueled by a new techno-arrogance, convinced that we now had the technology to produce any amount of fish.[3] The Canadian government was persuaded to fund the Salmonid Enhancement Program with a stated goal of doubling the catch of salmon in British Columbia. My job was to help make this program effective.

Some of the original results were indeed impressive. More than 15 percent of the fish released had survived to return to the hatcheries, or be caught. But, looking at the data over time, we see that survival rates kept declining, and it turned out that for some of the newly built hatcheries, it cost $1,000 to produce a single fish to be caught in the fishery. This was far from a good investment for the taxpayers of Canada. Over the years, I had become a serious critic of hatchery programs, especially in Canada, and had published many papers evaluating their performance. I was one of several people in the Pacific Northwest who showed that the optimism of the hatchery advocates had been thoroughly misplaced. Even though there was certainly a role for hatcheries in some places, they were not going to solve the problems of habitat loss and overfishing.

By 2000, Canadian salmon production had not doubled, but had sunk to one third of what it had been in the 1960s. The program was an abject failure, and the billion dollars spent had been a colossal waste.

In the United States, the Alaska hatchery program began in the mid-1970s, driven mostly by the decline in wild salmon. Alaska hatcheries now release about 1.4 billion juvenile salmon each year, mostly pink and chum that need little feeding before release. They achieve remarkable survival rates for such small fish, and there is no question that the total catch of salmon in the areas of major hatchery releases is higher than it would be without them.

Japan had maintained a hatchery program throughout much of the twentieth century, releasing several hundred million juvenile salmon. In the 1960s, their production ramped up to over 2 billion fish a year. The fish released were very small chum salmon that are cheap to raise, and about 40 million returning chum were caught each year. The Japanese hatchery program works well because there are almost no wild fish left; the cost of rearing chum is low; and there is no competition with wild fish for survival of the hatchery releases.

Japanese Marine Fish Enhancement

Japan maintains an extensive system of marine fish hatcheries primarily for coastal and salmon fisheries.[4] Around Japan's coast, thirty-four species of

A Closer Look

All regions showed rapid hatchery expansion, with a leveling off in the United States, Russia, and Japan, while Canadian production declined. Russian production declined just prior to the collapse of the Soviet Union and has rebuilt in the last 20 years. In the United States and Japan, many of the fisheries are largely dependent on hatchery production, while in Canada, the contribution to their fisheries was much less clear (see Figure 15.1).

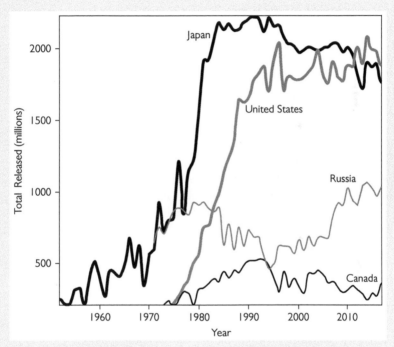

Figure 15.1 The number of juvenile Pacific salmon released from hatcheries in Japan, the United States, Canada, and Russia. Data from https://npafc.org/statistics/.

finfish, twelve species of crustaceans, twenty-five species of shellfish, and eight other species are enhanced primarily by hatchery operations. In 1994, 21 million flounder were released at 1,335 individual sites, as were 21 million red sea bream at 716 sites. Millions of other bream, herring, puffer fish, mud dabs, and others are also released. Also in 1994, 278 million juvenile prawns, 24 million shrimp, and 31 million swimming crabs were released. Add to that 10.9 billion seedlings of short-necked clams, 2.9 billion scallop seedlings, and of course the 2 billion salmon.

China's Marine Fish Enhancement

China has a marine enhancement program that rivals, and may exceed, that of Japan. As of 1999, at least forty-four species of marine fish had been successfully propagated, and for one species alone (large yellow croaker), over 300 million individuals from more than 200 individual hatcheries were released. Over 10 million individuals of six other species are released each year, plus at least a millions individuals of an additional six species. Between 2006 and 2010, about 7 billion juvenile fish were released each year. China also builds artificial reefs to enhance marine habitat, but there does not seem to be any significant evaluation of either biological or economic success of any of the enhancement programs.

Billions of Extra Fish: Is It Worth the Effort?

The history of stock enhancement and hatchery production is littered with failures that took decades to discover because of lack of any effective evaluation. It simply feels too good producing all those millions of fish to call it quits.

Ken Leber and Lee Blankenship, two biologists who work in marine fish enhancement, have led global efforts to make enhancement a scientific discipline that evaluates not only the biological and economic contribution of the released fish, but also searches for those practices to culture fish that are actually successful at enhancing wild production.[5] Through a series of scientific papers and conferences, they have led the way in making marine enhancement more effective. Their key points are:

- The need to identify quantitative measures of success.
- The need to be able to identify released fish (through marking) and assess stocking effects.
- The need to use empirical data to identify optimum release strategies.

All too often enhancement efforts have measured success by the number of fish released rather than any contribution to catch—indeed hatchery managers have sometimes been paid bonuses based on the number or weight of fish released without regard to fish surviving to adulthood. Many enhancement efforts could not even have identified survival because the hatchery fish were not marked in any way and looked the same as wild fish. Marking is particularly important when the goal is to supplement wild production, since one of the worries is that hatchery fish will compete with the wild ones for food. Such competition from hatchery releases may well reduce the production of wild fish, in which case there would be little if any net gain.

Pacific salmon have been particularly well-studied, and how hatchery fish impact wild fish has been a major concern. Hatchery fish certainly do compete with wild fish of the same species for food; some species of salmon will eat other species; hatchery fish have been shown to bring diseases to the wild fish and can genetically degrade the fitness of wild fish.

Because of the many hatchery programs and very good wild production in many areas, it now appears that there are more salmon in the North Pacific than before commercial fishing began. Studies have shown that in years when pink salmon are at high abundance in the ocean, growth and survival of sockeye plummet. That, in turn raises the question of whether the carrying capacity of the Pacific Ocean for salmon might have been reached.

How Does Aquaculture Affect Capture Fisheries?

Aquaculture produces over 50 million tons of highly nutritious food every year. It employs millions of people and is vital to the food security of perhaps a billion people.

A Closer Look

While global capture fisheries landings have stagnated, aquaculture continues to grow rapidly. It now produces more food than do direct capture fisheries. This graph includes production of plants and pearls. Global aquaculture production of fish in 2010 was 80 million tons (see Figure 15.2).

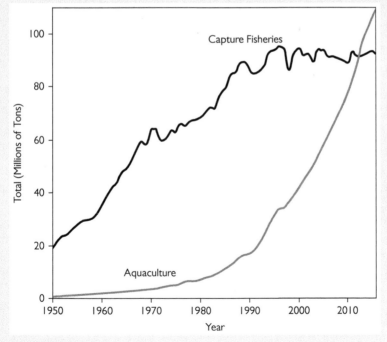

Figure 15.2 Global production of aquaculture (gray line) and capture fisheries landings (black line). Data from Food and Agriculture Organization of the United Nations.

Aquaculture began in the mists of time by simply capturing small fish and rearing them in ponds or any water where they could easily be harvested when grown. There is evidence that Aboriginal Australians captured eels and raised them in ponds 6,000 years ago. Chinese aquaculture goes back to at least 2,500 BC, and China is now the biggest producer of cultured fish. Almost all modern aquaculture began with taking wild fish, but over time, technology advanced so that hatchery-born fish can be used as the parents for the next generation.

An interesting exception to the usual forms of aquaculture is modern farming of southern bluefin tuna, where small tunas are captured off the coast of Australia and moved to large net pens where they are fed for 6 months to a year before being killed and sent to market. The timing of killing and marketing can be adjusted to get the best price. Since the number of bluefin taken for farming has a significant impact on the population, the catch for aquaculture is included as one component of the annual allowable catch.

Growing That Much Food in Confined Spaces Inevitably Comes with Its Own Challenges and Quite a Few Serious Problems

Aquaculture has become intertwined with capture fisheries in many ways, not all of them beneficial. Wild fish may be used as brood stock. When aquaculture fish escape into the wild, there are genetic repercussions on local wild stocks of the same species. When these escapees establish themselves, they become an often-unwelcome exotic species in the ecosystem. Diseases and parasites are transmitted to wild stocks. Wild fish are quite often used as feed for aquaculture fish. And there is the problem that mass produced fish can depress the market and price for wild fish.

Let Us Look at This in Detail

There will always be some fish that swim away from aquaculture facilities, particularly from marine net pens. When these facilities are within the spawning range of the same wild species, nature will win out, and interbreeding between aquaculture and wild fish can happen. To be successful, aquaculture operations select highly domesticated strains of a species that are genetically very different from their original wild species. Nevertheless, when they escape, they may interbreed with just that wild species. Farmed Atlantic salmon in Norway tend to escape in great numbers because of the vast scale of the operation (over 1 million tons produced), and interbreed with wild Atlantic salmon. There is much concern with the genetic impact on the wild fish, and it is often cited as a reason for the decline of wild Atlantic salmon in Norway.

Escapement from aquaculture operations can also establish nonnative species in new areas. Pacific oysters, once confined to the eastern coast of Asia, have spread to North America, Europe, Australia, and New Zealand. Farmed Pacific salmon in

Chile have escaped their pens and are sustaining wild breeding populations along the southern Chilean coast and into southern Argentina. Dozens of escaped species have become self-sustaining around the world, sometimes even competing with or preying on local species. In Puget Sound, 300,000 mature Atlantic salmon preferred the freedom of the open seas over daily feedings when their net pens failed in the summer of 2017, which quickly led to legislation banning Atlantic salmon net pen culture in Washington State.

Because of high densities, aquaculture operations are often ideal breeding sites for diseases and parasites that can then spread to local wild populations. Sea-lice from salmon farms are a serious concern to both aquaculture operators and the surrounding communities who depend on wild salmon. In 2006, a herpes-like virus spread from an abalone farm in the State of Victoria in Australia and rapidly devastated wild abalone in the region.

As we saw in Chapter 11, wild small pelagic fishes are a major source of fishmeal and fish oil used in aquaculture (and livestock) feeds. It is possible that the demand for aquaculture feed is driving the overexploitation of the small pelagics in some areas and, at the same time, takes away low-cost fish from poor people to fatten up high-value fish for the rich.

The massive Peruvian anchoveta fishery, on the other hand, is entirely based on the value of fishmeal and oil, because not very many people actually want to eat anchoveta.

But Can We Definitively Say That the Demand for Aquaculture Leads to Overexploitation?

This really depends on the management system. The status of these stocks is on average good: the big forage fisheries of the world are not overexploited (discussed further in Chapter 11). It is probably safe to say that if there were no markets for fishmeal and fish oil, the forage fish stocks would probably be fished much less, but at the global scale, there is no evidence that forage fish are being systematically overexploited because of aquaculture demand.

So Does Aquaculture Divert Fish as Food from the Poor to the Rich?

Probably yes in some cases, but we do not really know how common this is. Fishmeal and oil are made from a number of sources that do not compete with human food, including by-products of fish processing (heads, guts, frames, and skin), unmarketable low-quality fish, and fish that people do not seem to want, like anchoveta. It would definitely be useful to understand how much food for the poor is being diverted to aquaculture—and how much of aquaculture production is affordable for poor people, and how much is aimed at the rich.

Finally, aquaculture influences the price of wild capture fish. Between the late 1980s and early 2000s, the Bristol Bay salmon industry saw the price of salmon drop from over $2.00 a pound to less than 50 cents, due almost entirely to the great increase in global farmed salmon and trout. As opposed to wild capture fisheries, aquaculture can produce fresh fish consistently throughout the year with year-to-year stability. In some areas, wild-caught fish have definitely been squeezed out of the market. When catches of rockfish off the West Coast of the United States were reduced to rebuild some overexploited species, food services switched to farmed tilapia and catfish from Asia. Despite newly increased quotas, the market for rockfish remains collapsed, and it will take a sophisticated marketing strategy to lure consumers back to buying it.

. . . .

FURTHER READING

An overview of marine fish enhancement. Blaxter J. 2000. The enhancement of marine fish stocks. Advances in Marine Biology 38: 1–54.

A review of global aquaculture. Bostock J, McAndrew B, Richards R, et al. 2010. Aquaculture: Global status and trends. Philosophical Transactions of the Royal Society of London B: Biological Sciences 365: 2897–912.

An overview of Japanese enhancement. Masuda R and Tsukamoto K. 1998. Stock enhancement in Japan: Review and perspective. Bulletin of Marine Science 62: 337–58.

A review of hatcheries for salmonids. Naish K A, Taylor J E I, Levin P S, et al. 2008. An evaluation of the effects of conservation and fishery enhancement hatcheries on wild populations of salmon. Advances in Marine Biology 53: 61–194.

A review of Chinese management and the role of hatcheries. Shen G and Heino M. 2014. An overview of marine fisheries management in China. Marine Policy 44: 265–72.

A review of marine fish enhancement in China. Hong W-S and Zhang Q-Y. 2002. Artificial propagation and breeding of marine fish in China. Chinese Journal of Oceanology and Limnology 20: 41–51.

CHAPTER 16

Climate Change

Change in the Ocean: What Has Happened, What Will Happen?

The oceans are changing. We know that they are getting warmer (see Figure 16.1) and more acidic, and we know that this will greatly change the established patterns of how we get food from the sea and, consequently, how we manage fisheries.

We also know that we, the people, with our stupendously fertile minds, are the cause of the profound changes happening to our planet, and we can only hope that our collective minds will be nimble enough to adapt in good time to avoid or mitigate the worst.

Prediction Is Difficult, Especially about the Future (Niels Bohr)

Over the last several decades, intense scientific efforts have both documented historical changes in climate and attempted to forecast what that means for our future. I admit to being wary about our ability to predict the future. Our group at the University of Washington has been forecasting salmon returns for Bristol Bay for the next year only. We have been right within 20 percent in our 1-year forecasts, but we would never presume to attempt a 5-year prediction. This is not something we can do.

Weather and climate are relatively simple physical processes. Forecasts have become very reliable in recent years owing to better data collection and computational power. Marine ecosystems, however, are much more complex. Our forecasts concerning marine ecosystem changes for one relatively simple process like the salmon returns to Bristol Bay are much less reliable. But we can say with a great deal of certainty that the oceans have been getting warmer and more acidic. Right now, warming happens at warp speed in the Arctic, but since so few people live there, the alarm bells are not being heard by all yet. Even in the tropics, warming is evident. The physics and chemistry of ocean acidification, on the other hand, are very simple and highly predictable. More CO_2 in the atmosphere means more acidic oceans.

Ocean Recovery: a sustainable future for global fisheries? Ray Hilborn and Ulrike Hilborn, Oxford University Press (2019). © Ray Hilborn and Ulrike Hilborn 2019. DOI: 10.1093/oso/9780198839767.001.0001

It Is Already Happening

The most publicized changes in marine ecosystems associated with ocean warming have been coral-bleaching events. Bleaching happens when the corals expel the symbiotic algae as the water gets too hot, although other stressors can contribute. The corals rely on their algae for most of their energy, and if the algae do not manage to re-establish themselves, the corals die. The future is pretty dire for corals. There are estimates from 2016 that 31 percent of tropical corals worldwide were already bleached, and that at current rates, most corals will be dead within decades.

The biggest potential threat to marine fisheries is acidification due to ever-increasing CO_2 in the atmosphere. Acid dissolves calcium carbonate. This is why you can dissolve an eggshell in vinegar. Calcium carbonate is essential to making corals and shells precisely because it is insoluble in water—but not in acid.

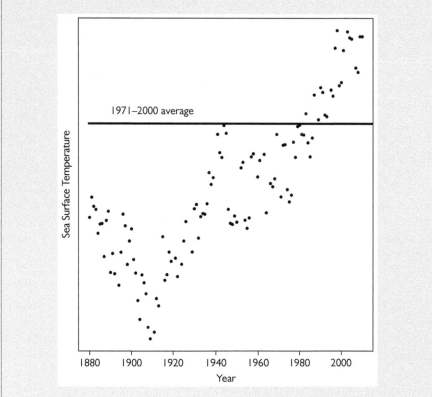

Figure 16.1 The average ocean temperature relative to the 1971–2000 average. Data from ftp://aftp.cmdl.noaa.gov/products/trends/co2/co2_mm_mlo.txt.

There is no question that the oceans are getting warmer.

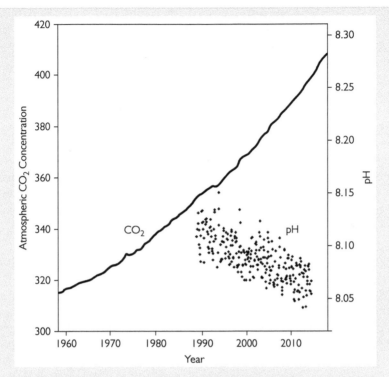

Figure 16.2 The atmospheric CO_2 concentration (solid line) and ocean acidity (Measured as pH). Data from https://www.ncdc.noaa.gov/data-access/marineocean-data/extended-reconstructed-sea-surface-temperature-ersst-v3b.

Again, there is no question that atmospheric CO_2 is rising, and as a result, the ocean is becoming more acidic.

Consequently, many marine organisms, from corals to oysters to tiny coccolitho-phores cannot make their shells when ocean acidity gets too high. Oyster growers in Washington State have already found the acidity of the ocean too high to suc-cessfully hatch their oysters and have had to move their operations to less acidic waters. We had better start imagining an ocean without the calcium-based shells that so many marine creatures depend on. Because coccolithophores and their photosynthesis form an important part of marine food chains, we may have to contemplate the entire food web of the oceans' fundamental transformation.

Ocean warming has already noticeably changed the life histories of marine fish. Quite a few species have found their usual habitat too warm and have moved toward the cooler polar regions. Taking data from scientific surveys of distribution of fish species from 1968 to 2011, a 2013 study[1] showed that fish are steadily moving toward the poles and to deeper and cooler water at rates roughly proportional to the increase in temperature. In the Northern Hemisphere, this is fairly evident already. Lobster in southern New England, salmon in California, and cod in the Gulf of Maine, all at the southern end of their historic ranges, are getting scarce. In Europe, movement of many stocks north of their traditional range is upsetting traditional allocation of catch between countries. Mackerel have moved to Iceland, but Iceland has little quota to catch them.

Living with Uncertainty

All scientific studies suggest that the oceans will continue to grow warmer and more acidic in the next decades. Moreover, there are estimates that there will be more severe storms, ocean levels will rise, and ocean oxygen concentration will decline. Even more alarming is the possibility that major ocean circulation patterns may disrupt the highly productive regions where upwelling brings nutrients from the deep waters of the ocean, and that large-scale phenomena like the Gulf Stream, that warms the northern parts of Europe, may shift. The range of possible changes is terrifying.

If we look at the range in climate forecasts that come from the Intergovernmental Panel for Climate Change, the uncertainty is large, because we do not know what our future emissions of greenhouse gases will be. If we add the even greater uncertainty of how ocean and freshwater ecosystems will respond, I conclude that forecasting changes provides at best a range of what the future might hold and that it is far more important to understand how societies can adapt to whatever changes will occur.

The Need for Adaptation

Adaptation is the name of the game. How will individual species and ecosystems adapt? Will they? Can they even?

Take the tiny coccolithophores, so important to marine photosynthesis, that come in a great many varieties. Some of them are more resilient than others to changes in ocean acidity. It is possible that the species mix of coccolithophores will change, but the group, as a whole, will retain its ecosystem function. Of course, there will be very strong natural selection for individuals that are resistant to acidity changes in any species, and we expect to see evolution happen at a rapid pace. But the uncertainties remain whether species can adapt fast enough and where the chemical and physical limits to such speedy adaptation are.

Whatever Happens, Some Will Win, Some Will Lose

Fisheries management systems are able to respond to climate-induced changes in fish productivity and abundance.[2] They can reduce fishing pressure on species that become less productive, and increase it on fisheries that become more productive.

Wherever management systems track productivity and adjust fishing pressure, this kind of adaptation will happen without any changes to the system. Fishermen will naturally adapt by moving to where the fish are.

But wherever fish have been allocated to different fishing fleets by both species and area, there will be winners and losers. This is a serious problem in Atlantic Europe, where quotas were allocated to different countries based on fish distributions 30 years ago. Fish are notoriously unmindful of boundaries, and as some species have moved farther north, a country may find a new species within its fisheries but no quota to catch it. There are already eighteen species, including especially desirable cod and mackerel, that have shifted north in Atlantic Europe, leading to a severe mismatch between traditional allocations and what turns up in the nets. The current EU system will adjust the total quota appropriately as stock abundance and productivity change, but there are currently no effective provisions for reallocating quota to individual countries.

Can We Manage the New Normal with the Old Rules?

Over the last 50 years, we have seen many efforts to reduce fishing pressure that have unfortunately reduced the fisheries' ability to adapt quickly and meaningfully. Perhaps the most problematic of these are license limitation and limited entry because they prohibit switching between fishing opportunities. Historically, fishermen could switch between species in their area whenever relative abundance changed, and they can still do so in many fisheries around the world where entry is not limited. Newfoundland fishermen could traditionally fish for a wide range of species including cod, lobster, shrimp, scallop, and seals, depending on what was abundant at the time. When limited entry was introduced to keep the growth of fishing capacity under control, licenses were given for one or two individual fisheries only. No more switching. Consequently, when groundfish collapsed and the invertebrates boomed, the cod fishermen went on unemployment assistance, and the shrimp, scallop, crab, and lobster fishermen got rich.

Even within the same species, we commonly allocate fish by specific region. As the fish move toward the poles, the fishermen there will benefit while those closer to the tropics will catch fewer. Fisheries with area-specific allocation are also denied the opportunity to adapt.

I do not have specific solutions for how to modify current fish allocation to make it more adaptable, but adapt it must, and the subject needs to go to the top of any fisheries management agenda.

Scientists Too Will Have to Rethink Stock Assessments

The same is true for science. Traditional stock assessments have largely been based on the assumption that biological processes such as births, deaths, and somatic growth rates will be the same in the future as in the past. We know that mortality rates from predation depend on how many predators there are, and that growth rates change with competition. We also know that many stocks show major changes in recruitment. Yet most stock assessments still assume that average natural mortality, somatic growth, and recruitment are and will be the same. It is paramount that we begin to allow for systematic changes in these important biological processes because management reference points, such as the harvest rate that produces long-term maximum yield, will be changing over time.

The state of the art method to evaluate management policies now used in many fisheries is management strategy evaluation (MSE). In essence, MSE looks for management rules that provide good outcomes across the entire range of possible fishery and ecosystem dynamics. In the face of climate change, MSE is even more important, but, urgently, it needs to be expanded to look at not only how allowable catches are determined, but also, explicitly, the allocation and dynamics of fishing fleets and their communities.

Suggestions for Adaptation

While the United States is officially still unmindful of climate change at the time I write this, much work is being done on how we might best cope. Suggestions on how we adapt for fisheries and aquaculture include:[2]

- A good understanding of the ecological functions, socioeconomics, and institutional contexts of a given fishery should be required to strengthen resilience and devise the most appropriate adaptation response.
- Adaptation should be viewed as an on-going and iterative process, incorporating flexibility and feedback to learn from experiences and avert new risks.
- Climate change adaptation should start with an accurate assessment of current climate variability and should consider future climate changes as prerequisites for determining early low- or no-regret options.
- It is important to consider transboundary issues.
- Evaluations of success are often missing from adaptation studies.

Let us hope that wisdom and luck will be with us.

FURTHER READING

An analysis of the impact of climate change on the fisheries of Bangladesh. Allison E H, Perry A L, Badjeck M C, Neil Adger W, Brown K, Conway D, Halls A S, et al. 2009. Vulnerability of national economies to the impacts of climate change on fisheries. Fish and Fisheries 10: 173–96.

An analysis of fish distribution shifts in the Northeast Atlantic. ICES. 2917. Report of the Working Group on Fish Distribution Shifts (WKFISHDISH), November 22–25, 2016, ICES HQ, Copenhagen, Denmark. ICES Document ICES CM 2016/ACOM: 55, 197.

A look at northward shifts in the distribution of fish. Pinsky M L, Worm B, Fogarty M J, Sarmiento J L, and Levin S A. 2013. Marine taxa track local climate velocities. Science 341: 1239–42.

CHAPTER 17

The Future of Fisheries

The Oceans Will Continue to Supply Food to the World

It should be clear by now that I am optimistic about the future of fisheries to supply food to the world. As we saw in Chapter 5, most assessed and managed fisheries are increasing or fluctuating around desired levels. This includes almost all the major fisheries of North America, Europe, Russia, and Japan; many of the large fisheries of Latin America; most major fisheries of South Africa, New Zealand, and Australia; and global tuna fisheries. All of these account for roughly half of global fish production. As long as the management systems stay in place and the marine ecosystems remain productive, the fish they produce can be sustainably harvested and contribute to food security.

The other half of the world's fisheries, those that are largely unmonitored and unmanaged, are the concern. Some may continue to deteriorate if fishing pressure stays too high; others may remain in an overfished condition, but continue to produce food, just not as much as they could if well-managed. One study attempted to forecast the future of global fisheries if well-managed fisheries continue to be well-managed, and unmanaged fisheries become so overfished that fishing pressure ceases to grow.[1] Surprisingly, they forecast only a slight difference in global catch. A future of the seas being empty of fish is just not what has been seen. While it is true that the unmanaged fisheries will have much lower fish abundance and most of their large fish will have almost disappeared, lots of fish will still be caught. Most forecasts suggest that even without good management, food will still come from the sea—but it will be distinctly less, and in some cases, dramatically less than could be achieved with good management.

Feeding the Teeming Billions

Our world will get more crowded. Barring a universal global catastrophe, more people will have more disposable income and will want to eat better. That and

Ocean Recovery: a sustainable future for global fisheries? Ray Hilborn and Ulrike Hilborn, Oxford University Press (2019). © Ray Hilborn and Ulrike Hilborn 2019. DOI: 10.1093/oso/9780198839767.001.0001

the inevitable rise in the world's population will put severe demands on the supply of fish.

Key projections suggest that by 2050 there will be 2.3 billion more of us. Consequently, the demand for fish protein will rise sharply in all the regions where fisheries are most important and least managed. Moreover, annual GDP in East Asia will probably grow by 5.4 percent, and in South Asia, by 4.0 percent, through 2030, putting unprecedented pressure on fisheries for more and better seafood.[2]

We must be prepared to at least maintain, and possibly expand, both capture fisheries and aquaculture to meet the coming intense demand.

All Things May Not Remain Equal

This optimism assumes that the oceans remain productive, and this is not guaranteed. Global warming and ocean acidification are always lurking in the shadows to disrupt the ability of the seas to provide food. The oceans may become less productive for our target species, and food webs may change so that the energy from the sun flows through the food web in different ways, producing lots of fish—but perhaps not the kind of fish we prefer to eat. We simply do not know.

Regardless of whether we are thinking of small-scale or large-scale, freshwater or marine, commercial or recreational fisheries, the key is being able to adapt to future changes. The adaptation has to be in how we assess and regulate our fisheries, and how our fishing industries are structured.

There is the potential to increase food from the sea—perhaps dramatically. Most of the species in marine ecosystems are unexploited. If we wanted to harvest lower on the food chain, more food could be produced. The most obvious source is Antarctic krill—the small shrimp-like invertebrates that are the primary food of Antarctic whales. Krill is being harvested in limited quantities and is used as a human food, a food supplement, and in aquaculture. Estimates of the potential yield of krill in the Antarctic are up to tens of millions of tons. The actual harvest of Antarctic krill has been limited both by weak markets for the product and concern about the impact of krill harvesting on whales and seals.

There is a large, possibly very large, abundance of fish that live in the high seas at depths of 200 to 1000 meters in what is known as the meso-pelagic zone of the ocean.[3] Estimates range from 1 billion to 10 billion tons. For comparison, the abundance of all of the species exploited now is thought to be on the order of only 1 billion tons. If these meso-pelagic fishes could be captured in an economically viable way, it is possible that ocean fish production could be doubled. The technology is not on the immediate horizon, but the potential alone speaks loudly that, despite popular perception, most of the abundance of fish and invertebrates in the sea remains untouched.

Notwithstanding the potential of the oceans to continue to produce food, challenges abound.

Small-Scale Fisheries

Fisheries that use small boats or other fishing gear typically fish close to shore and are local, often tied to specific communities and surrounding waters. This would include everything from dugout canoes in West Africa to the fleets of small boats in many European waters. The main challenge is management. It is not possible to implement the kind of data collection, analysis, and regulations that characterize larger scale fisheries management. Different systems are needed, as we discussed extensively in Chapter 3. At present, the majority of them are not managed very intensively, and achieving their potential is going to require finding ways to do it effectively.

Being local and generally confined to the near shore, small-scale fisheries are particularly vulnerable to changes such as erosion, loss of wetlands, and land-based pollution. As inland waters, they are even more vulnerable to dams, pollution, and water diversion.

Geography hamstrings them because they have fewer options to respond to climate-induced changes in the distribution of fish stocks. If their target species swim away toward the poles, they cannot follow as easily as the larger, more mobile fishing boats can.

To be successful, they need legally or culturally defined exclusive access, both to exclude large-scale fisheries and to provide for successful local management. If they must compete with bigger boats or with neighboring communities, they are much less likely to be able to manage their resources effectively.

Another major challenge is economic—they have to compete with large-scale fisheries and with aquaculture. On the whole, larger vessels are economically more efficient. Fish is the most traded food product, and prices depend on global production. If an industrial fleet or aquaculture in another country can produce a similar product more cheaply, that will set the global price. We saw in Chapter 1 that much of the small-boat fleet fishing salmon in Bristol Bay was driven to financial collapse by the rapid growth in salmon aquaculture far away in Norway and Chile.

Where access to fisheries is a tradeable commodity, either through transferable limited-entry permits or transferable quotas, there is a near universal tendency toward consolidation in larger vessels and loss of fishing permits for the small ones. This can be due to the economic efficiency of larger vessels or because bigger operators have better access to capital. Wherever the catch of small-scale fisheries is not protected by law or custom, or small-scale fisheries have an economic advantage over larger vessels, I think they may well mostly diminish.

Recreational Fisheries

This is also a small-scale fishery with similar problems of science, management, and enforcement. Because marine anglers are also dependent on nearshore resources, the fish they want to catch are very sensitive to loss of wetland habitat and terrestrial pollution in all its permutations.

The anglers' strength is in their numbers. They may be widely dispersed in area and country, but one thing unites them—commercial fishing is their competition. The battles over catch allocation are fought with conviction and the ballot box. Political pressure is definitely working in the countries I know best. In Canada, the United States, Australia, and New Zealand, anglers get what they want—the larger share of their desired species, and this is not likely to change.

Flexing their muscles getting their share of the catch, anglers are sometimes recalcitrant to recognize the need for regulations. They argue that each one of them usually catches not that many fish, not wanting to admit or even countenance the idea that collectively they too can overexploit resources and need restrictions. There is an ingrained habit wherever sport fishermen have to share the catch with their commercial brethren to blame them for any problems.

When I worked on the Chinook salmon fishery in British Columbia in the early 1980s, the average number of fish caught per year per angler was one. Given the impressive numbers of anglers and the economic importance of the recreational fishery, there just are not enough fish available to make everyone happy. Wildlife managers issue tags to deal with scarcity of the resource—each licensed hunter might be able to kill one or two of his desired animal per season. If you want to kill something really rare, you will have to enter a lottery to get a tag. The anglers, too, will eventually face this kind of limit in some marine fisheries, but so far, the mere idea of tags is heresy.

Industrial Fisheries

The challenges faced by industrial fisheries depend greatly on the location. The large industrial fisheries of much of Asia, Africa, and some of Latin America still have not developed effective management systems, which prevents them from achieving their potential.

In the countries with effective management, abundance is increasing, but catch is not. In the United States, Canada, Europe, New Zealand, and Australia, the battle is over trade-offs between producing food or generating jobs and preserving marine ecosystems in an undisturbed condition. Active and well-funded non-governmental organizations (NGOs) want to limit fishing by establishing large marine protected areas, banning bottom trawling, and severely restricting catch limits. In Australia and New Zealand, where environmental NGOs have been par-ticularly effective, I fear that large-scale fisheries are in danger of losing their access to fish.

Who Owns the Oceans?

Right now, our oceans are up for grab. Competing demands for wind farms, seabed mining, and oil exploration are on track to push fishing out of its traditional fishing grounds. The fishing industry is rightly worried that with the unwitting help of powerful NGOs, these interests will pursue an end to fishing so that oil, mining, and energy production can advance unfettered. Given that some of the environmental NGOs' money comes from oil is not only ironic but should definitely be cause for concern.

Only by demonstrating sustainability and low environmental impact can commercial fisheries get the public on their side. Therefore, my advice to commercial fisheries is to get on with it—install around-the-clock monitoring by observers or cameras. Accept continuous and public tracking of vessels. Keep working on modifying gear to reduce and ideally eliminate any impacts on nontarget species and vulnerable ecosystems. Be transparent and get the word out about how important you are as stewards of the world's food security—or you will get pushed out of existence.

Closing Thoughts

The evidence is strong that ocean fishing is one of the most environmentally friendly ways to produce food, and it is an important part of global food security. Capture fisheries can be and are now largely sustainably managed. But across all dimensions, we can and must do better. Unmanaged fisheries can be managed; impacts of fishing on marine ecosystems can be reduced; and illegal fishing can be largely eliminated. There is considerable room for increasing food production from the fish we catch already by using all of it—heads, tails, guts, and skeleton can all be turned into fishmeal and fish oil.

In the End, We, the People, Will Have to Make the Tough Choices

As fisheries management improves, all societies will face difficult trade-offs. Do we want to maintain small-scale fishing communities? Should recreational fisheries grow and eliminate commercial fishing? And how much impact on marine ecosystems do we want to allow?

Different countries will make different choices, but it is high time for all of us to start talking about what it is we want our fisheries to do and how much value we assign to their existence.

Let us just hope that we, the people, will not disappoint us.

ENDNOTES

CHAPTER 1

1. Schindler D E, Leavitt P R, Brock C S, et al. 2005. Marine-derived nutrients, commercial fisheries, and production of salmon and lake algae in Alaska. Ecology 86: 3225–31.
2. Schindler D E, Hilborn R, Chasco B, et al. 2010. Population diversity and the portfolio effect in an exploited species. Nature 465: 609–13.

CHAPTER 2

1. Brundtland G, Khalid M, Agnelli S, et al. 1987. Our common future: Report of the 1987 World Commission on Environment and Development. United Nations, Oslo, 59.
2. Matson P, Clark W C, and Andersson K. 2016. Pursuing sustainability: A guide to the science and practice. Princeton: Princeton University Press.
3. Larkin P A. 1977. An epitaph for the concept of maximum sustained yield. Trans Am Fish Soc 106: 1–11.

CHAPTER 3

1. Johannes R E. 1981. The words of the lagoon: Fishing and marine lore in the Palau District of Micronesia. Berkeley: University of California Press
2. Johannes R E. 1978. Traditional marine conservation methods in Oceania and their demise. Annu Rev Ecol Syst 9: 349–64.
3. Hilborn R. Marine biota. In: Turner BL, III (ed.). 1990. The earth as transformed by human action. Cambridge: Cambridge University Press, 371–86.
4. Huxley T H. 1884. Inaugural address. Fisheries Exhibition Literature 4: 1–22.
5. Hannesson R, Salvanes K G, and Squires D. 2008. Technological change and the tragedy of the commons: The Lofoten fishery over hundred and thirty years. Land Economics 86.4: 746–65.
6. Ostrom E. 1990. Governing the commons: The evolution of institutions for collective action. Governing the commons: The evolution of institutions for collective action. Cambridge: Cambridge University Press.
7. http://jgshepherd.com/thoughts/.
8. Melnychuk M, Peterson, E., Elliott, M., and Hilborn, R. 2017. Fisheries management impacts on target species status Proceedings of the National Academy of Sciences 114: 178–83.

CHAPTER 4

1. Hardin G. 1968. The tragedy of the commons: The population problem has no technical solution; it requires a fundamental extension in morality. Science 162: 1243–8.
2. Ibid.

3. Ostrom E. 2008. Tragedy of the commons. In: S N Durlauf and E L Blume (eds), The New Palgrave dictionary of economics, 2nd edition. Basingstoke: Palgrave MacMillan.

4. Adams D J. 2012. Four thousand hooks: A true story of fishing and coming of age on the high seas of Alaska. Seattle: University of Washington Press.

5. Costello C, Gaines S D, and Lynham J. 2008. Can catch shares prevent fisheries collapse. Science 321: 1678–81.

6. Essington T E, Melnychuk M C, Branch T A, et al. 2012. Catch shares, fisheries, and ecological stewardship: A comparative analysis of resource responses to a rights-based policy instrument. Conservation Letters 5: 186–95.

CHAPTER 5

1. Worm B, Barbier E B, Beaumont N, et al. 2006. Impacts of biodiversity loss on ocean ecosystem services. Science 314: 787–90.

2. Ludwig D, Hilborn R, and Walters C. 1993. Uncertainty, resource exploitation, and conservation: Lessons from history. Science 260: 17–36.

3. Kurlansky M. 1997. Cod: A biography of the fish that changed the world. New York: Walker and Co.

4. Worm B, Hilborn R, Baum J K, et al. 2009. Rebuilding global fisheries. Science 325: 578–85.

5. Stokstad E. 2009. Détente in the fisheries war. American Association for the Advancement of Science 324: 170–1.

6. www.ramlegacy.org.

7. Myers R A and Worm B. 2003. Rapid worldwide depletion of predatory fish communities. Nature 423: 280–3.

8. Walters C J. 2003. Folly and fantasy in the analysis of spatial catch rate data. Can J Fish Aquat Sci 60: 1433–6.

9. Hampton J, Sibert J R, Kleiber P, et al. 2005. Decline of pacific tuna populations exaggerated? Nature 434: E1–E2.

10. Juan-Jorda M S, Mosqueira I, Cooper AB, et al. 2011. Global population trajectories of tunas and their relatives. Proceedings of the National Academy of Sciences USA 108: 20650–5.

11. Melnychuk M, Peterson E, Elliott M, and Hilborn R. 2017. Fisheries management impacts on target species status Proceedings of the National Academy of Sciences 114: 178–83.

12. Szuwalski C S, Burgess M G, Costello C, et al. 2016. High fishery catches through trophic cascades in china. Proceedings of the National Academy of Sciences 114(4): 717–21.

13. Pauly D, Christensen V, Dlasgaard J, et al. 1998. Fishing down marine food webs. Science 279: 860–3.

14. Branch T A, Watson R, Fulton E A, et al. 2010. The trophic fingerprint of marine fisheries. Nature 468: 431–5.

15. Sethi S A, Branch T A, and Watson R. 2010. Global fishery development patterns are driven by profit but not trophic level. Proceedings of the National Academy of Sciences USA 107: 12163–7.

16. Christensen V, Coll M, Piroddi C, et al. 2014. A century of fish biomass decline in the ocean. Mar Ecol Prog Ser 512: 155–66.

CHAPTER 6

1. Tyedmers P H, Watson R, and Pauly D. 2005. Fueling global fishing fleets. Ambio 34: 635–8.

2. Steinfeld H, Gerber P, Wassenaar T, et al. 2006. Livestock's long shadow: Environmental issues and options. Rome: Food and Agriculture Organization of the United Nations.

3. Pelletier N, Audsley E, Brodt S, et al. 2011. Energy intensity of agriculture and food systems. Annu Rev Environ Resour 36: 223–46.

4. Hall S J, Delaporte A, Phillips M J, Beveridge M, and O'Keefe M. 2011. Blue frontiers: Managing the environmental cost of aquaculture. Penang, Malaysia: The World Fish Center.
5. Montgomery D R. 2007. Dirt: The erosion of civilizations. Berkeley: University of California Press.
6. Essington T E. 2006. Pelagic ecosystem response to a century of commercial fishing and whaling. Whales, whaling, and ocean ecosystems, 1st edn. Berkeley: University of California Press, 38–49.

CHAPTER 7

1. FAO. 2012. Technical Guidelines for Responsible Fisheries: Recreational Fisheries. No. 13. Rome: FAO.

CHAPTER 8

1. Lymer D, Marttin F, Marmulla G, and Bartley D. 2016. A global estimate of theoretical annual inland capture fisheries harvest. In: W W Taylor, D M Bartley, C I Goddard, et al. (eds), Freshwater, fish and the future: Proceedings of the global cross-sectoral conference. Rome: Food and Agriculture Organization of the United Nations; East Lansing: Michigan State University; Bethesda, MD: American Fisheries Society.
2. Dene-Hern Chen. 2018. Once written off for dead, the Aral Sea is now full of life. National Geographic (March). https://www.nationalgeographic.co.uk/environment-and-conservation/2018/03/once-written-dead-aral-sea-now-full-life.
3. Witte F, Goldschmidt T, Wanink J, van Oijen M, Goudswaard K, Witte-Maas E, and Bouton N, 1992. The destruction of an endemic species flock: quantitative data on the decline of the Haplochromine cichlids of Lake Victoria. Environ Biol Fish 34: 1–28.
4. Bennett E and Thorpe A. 2008. Review of river fisheries valuation in Central and South America. In: A E Neiland (ed), Tropical river fisheries valuation: Background papers to a global synthesis. Penang, Malaysia: The World Fish Center, 2–46.
5. https://www.popsci.com/environmental-damage-amazon-river-dams.
6. Taylor W W and Bartley D M. 2016. Call to action–The "Rome Declaration": Ten steps to responsible inland fisheries. Fisheries 41: 269–9.

CHAPTER 9

1. Hilborn R, Stewart I J, Branch T A, and Jensen O P. 2012. Defining trade-offs among conservation of species diversity abundances, profitability, and food security in the California Current bottom-trawl fishery. Conserv Biol 26: 257–66

CHAPTER 10

1. Watling L and Norse E A. 1998. Disturbance of the seabed by mobile fishing gear: A comparison to forest clearcutting. Conserv Biol 12: 1180–97
2. NRC. Effects of trawling and dredging on seafloor habitat. Washington, DC: National Academy Press, 2002.
3. Branch T A, Hilborn R, and Bogazzi E. 2005. Escaping the tyranny of the grid: a more realistic way of defining fishing opportunities. Can J Fish Aquat Sci 62: 631–42.
4. Rijnsdorp A, Bastardie F, Bolam S G, Buhl-Mortensen L, Eigaard O R, Hamon K, Hiddink J, Hintzen N T, Ivanović A, and Kenny A. 2015. Towards a framework for the quantitative assessment of trawling impact on the seabed and benthic ecosystem. ICES J Mar Sci 73: i127–i138.
5. Amoroso R O, Pitcher C R, Rijnsdorp A D, McConnaughey R A, Parma A M, Suuronen P, Eigaard O R, Bastardie F, Hintzen N T, Althaus F, and Baird S J. 2018. Bottom trawl fishing

footprints on the world's continental shelves. Proceedings of the National Academy of Sciences 115(43): E10275–85.

6. Hiddink J G, Jennings S, Sciberras M, Szostek C L, Hughes K M, Ellis N, Rijnsdorp A D, McConnaughey R A, Mazor T, and Hilborn R. 2017. Global analysis of depletion and recovery of seabed biota after bottom trawling disturbance. Proceedings of the National Academy of Sciences 114: 8301–6.

7. Collie J, Hiddink J G, Kooten T, Rijnsdorp A D, Kaiser M J, Jennings S, and Hilborn R. 2017. Indirect effects of bottom fishing on the productivity of marine fish. Fish Fish 18: 619–37.

CHAPTER 11

1. Schwartzlose R A, Alheit J, Bakun A, et al. 1999. Worldwide large-scale fluctuations of sardine and anchovy populations. S Afr J Mar Sci 21: 289–347.

2. Cushing D H. 1982. Climate and fisheries. London; New York: Academic Press.

3. Baumgartner T R, Soutar A, and Ferreira-Bartrina V. 1992. Reconstruction of the history of pacific sardine and northern anchovy populations over the past two millennia from sediments of the Santa Barbara Basin, California. CalCOFI Report 33: 24–40.

4. www.ramlegacy.org.

5. Pikitch E, Boersma P D, Boyd I, et al. 2012. Little fish, big impact: Managing a crucial link in ocean food webs. Washington, DC: Lenfest Ocean Program, 108.

6. Punt A, MacCall A D, Essington T E, Francis T B, Hurtado-Ferro F, Johnson K F, Kaplan I C, Koehn L E, Levin P S, and Sydeman W J. 2016. Exploring the implications of the harvest control rule for pacific sardine, accounting for predator dynamics: A mice model. Ecol Modell 337: 79–95.

CHAPTER 12

1. Melnychuk M, Peterson E, Elliott M, and Hilborn, R. 2017. Fisheries management impacts on target species status. Proceedings of the National Academy of Sciences 114: 178–83.

2. FAO's Voluntary Guidelines for Flag State Performance, adopted in 2014.

3. FAO. 2015. Report of the expert workshop to estimate the magnitude of illegal, unreported and unregulated fishing globally. Rome, February 2–4, 2015. FAO Fisheries Technical Report 1106: 38.

4. Polacheck T. 2012. Assessment of IUU fishing for southern bluefin tuna. Mar Policy 36: 1150–65.

CHAPTER 13

1. Cannon J, Sousa P, Katara I, et al. 2018. Fishery improvement projects: Performance over the past decade. Marine Policy https://www.sciencedirect.com/science/article/pii/S0308597X18300393.

2. California Environmental Associates. 2012. Charting a course to sustainable fisheries. https://www.ceaconsulting.com/casestudies/charting-a-course-to-sustainable-fisheries/.

3. Ovando D. 2018. Of fish and men: Using human behavior to improve marine resource management. PhD thesis. University of California Santa Barbara.

4. Moore PA. 2010. Confessions of a Greenpeace dropout: The making of a sensible environmentalist. Vancouver: Beatty Street Publications.

5. NRDC. https://www.nrdc.org/es/taxonomy/term/10429.

6. Greenpeace. https://www.greenpeace.org/usa/oceans/issues/overfishing-destructive-fishing/.

7. World Wildlife Fund. https://:wwf.panda.org/our_work/oceans/solutions/sustainable_fisheries/.

8. Oceana. https://oceana.org/our-campaigns/save_oceans_feed_world/campaign.

9. Conservation International. https://www.conservation.org/How/Pages/Transforming-wild-fisheries-and-fish-farming.aspx.

10. The Nature Conservancy. https://www.nature.org/en-us/about-us/where-we-work/united-states/oregon/stories-in-oregon/sustainable-fisheries/.

11. Environmental Defense Fund. https://www.edf.org/oceans.

12. Tracey S, Buxton C, Gardner C, et al. 2013. Super trawler scuppered in Australian fisheries management reform. Fisheries 38: 345–50.

13. Hughey K F, Kerr G N, and Cullen R. 2016. Public perceptions of New Zealand's environment: 2016. Christchurch, New Zealand: EOS Ecology.

14. Hilborn R. 2018. Losing grounds: Self-report or report by force. National Fisherman October 2018: 6–8.

15. Wilson T. 2012. Naked extortion? Environmental NGOs imposing [in] voluntary regulations on consumers and business. Journal of Oil Palm, Environment and Health 3: 16–29.

CHAPTER 14

1. Pikitch E K, Santora C, Babcock E A, Bakun A, Bonfil R, Conover D O, Dayton P, et al. 2004. Ecosystem-based fishery management. Science 305: 346–7.

2. Hilborn R. 2011. Future directions in ecosystem based fisheries management: A personal perspective. Fisheries Research 108: 235–9.

3. Trochta J T, Pons M, Rudd M B, Krigbaum M, Tanz A, and Hilborn R. 2018. Ecosystem-based fisheries management: Perception on definitions, implementations, and aspirations. PLoS ONE 13: e0190467.

4. Lubchenco J, Gaines S, Warner R, Airame S, and Simler B. 2002. Partnership for Interdisciplinary Studies of Coastal Oceans. The Science of Marine Reserves http://www.piscoweb.org.

CHAPTER 15

1. Taylor III J E. 2009. Making salmon: An environmental history of the Northwest fisheries crisis. Seattle: University of Washington Press.

2. Hilborn R and Winton J. 1993. Learning to enhance salmon production—Lessons from the Salmonid Enhancement Program. Canadian Journal of Fisheries and Aquatic Sciences 50: 2043–56.

3. Larkin P A. 1974. Play it again Sam—An essay on salmon enhancement. Journal of the Fisheries Research Board of Canada 31: 1433–59.

4. Kitada S and Kishino H. 2006. Lessons learned from Japanese marine finfish stock enhancement programmes. Fisheries Research 80: 101–12.

5. Blankenship H L and Leber K M. 1995. A responsible approach to marine stock enhancement. In American Fisheries Society Symposium No. 15, Bethesda, MD: 167–75.

CHAPTER 16

1. Pinsky M L, Worm B, Fogarty M J, Sarmiento J L, and Levin S A. 2013. Marine taxa track local climate velocities. Science 341: 1239–42.

2. Food and Agriculture Organization of the United Nations. 2018. Impacts of climate change on fisheries and aquaculture: Synthesis of current knowledge, adaptation and mitigation options. FAO Fisheries and Aquaculture Technical Paper No. 627.

CHAPTER 17

1. Costello C, Ovando D, Clavell, T, Strauss C K, Hilborn R, Melnychuk M C, Branch T A, et al. 2016. Global fishery prospects under contrasting management regimes. Proceedings of the National Academy of Sciences 113: 5125–9.

2. Vannuccini S, Kavallari A, Bellù L G, et al. 2018. Understanding the impacts of climate change for fisheries and aquaculture: Global and regional supply and demand trends and prospects: Synthesis of current knowledge, adaptation and mitigation options. In: Barange M, Bahri A, Beveredge M C M, et al. (eds). Impacts of climate change on fisheries and aquaculture. Rome: Food and Agriculture Organization of the United Nations.
3. Kaartvedt S, Staby A, and Aksnes D L. 2012. Efficient trawl avoidance by mesopelagic fishes causes large underestimation of their biomass. Mar Ecol Prog Ser 456: 1–6.

INDEX

Tables and figures are indicated by an italic *t* and *f* following the page number.